Getting Started with
M A T L A B

A Quick Introduction for
Scientists and Engineers

RUDRA PRATAP

Department of Mechanical Engineering
Indian Institute of Science, Bangalore

Sibley School of Mechanical and Aerospace Engineering
Cornell University

New York · Oxford
OXFORD UNIVERSITY PRESS
2002

Oxford University Press

Oxford   New York
Athens Auckland Bangkok Bogotá Buenos Aires Cape Town
Chennai Dar es Salaam Delhi Florence Hong Kong Istanbul Karachi
Kolkata Kuala Lumpur Madrid Melbourne Mexico City Mumbai Nairobi
Paris São Paulo Shanghai Singapore Taipei Tokyo Toronto Warsaw

and associated companies in
Berlin   Ibadan

Copyright © 2002 by Oxford University Press, Inc.

Published by Oxford University Press, Inc.
198 Madison Avenue, New York, New York, 10016
http://www.oup-usa.org

Oxford is a registered trademark of Oxford University Press

ISBN: 0-19-515014-7

Printing number: 9 8 7 6 5 4 3 2

Printed in the United States of America
on acid-free paper

Getting Started with

M A T L A B

A Quick Introduction for Scientists and Engineers

Books are to be returned on or before
the last date below.

26 SEP 2003

- 8 DEC 2003

LIBREX —

Disclaimer
Under no circumstances
does the author assume any
responsibility and liability thereof, for any
injury caused to the reader by toxic fumes and
explosions resulting from mixing incompatible
matrices and vectors. Array operations are known to
cause irritability and minor itching to beginners. The
author, however, might buy the reader a cup of coffee
in the case of serious distress. In rare cases of very
flattering comments or very creative suggestions
about improving this book, the author might even
buy the reader lunch. The reader is encouraged
to try his/her luck by sending comments
to pratap@mecheng.iisc.ernet.in or
rp28@cornell.edu

To Ma Gayatri

and my parents
Shri Chandrama Singh and Smt. Bachcha Singh

Contents

Preface

I enjoy MATLAB, and I want you to enjoy it too—that is the singular motivation behind this book. The first and foremost goal of this book is to get you started in MATLAB quickly and pleasantly.

Learning MATLAB changed the meaning of scientific computing for me. I used to think in terms of machine-specific compilers and tables of numbers as output. Now, I expect and enjoy interactive calculation, programming, graphics, animation, and complete portability across platforms—all under one roof. MATLAB is simple, powerful, and for most purposes quite fast. This is not to say that MATLAB is free of quirks and annoyances. It is not a complete miracle drug, but I like it and I think you will probably like it too.

I first used MATLAB in 1988 in a course on matrix computation taught by Tom Coleman at Cornell University. We used the original 1984 commercial version of MATLAB. Although the graphics capability was limited to bare-bones 2-D plots, and programming was not possible on the mainframe VAX, I still loved it. After that I used MATLAB in every course I took. I did all the computations for my Ph.D. dissertation in nonlinear dynamics using MATLAB. Since then I have used MATLAB in every engineering and mathematics course I have taught. I have enthusiastically tried to teach MATLAB to my friends, colleagues, students, and my 4-year-old daughter. I have given several introductory lectures, demonstrations, and hands-on workshops. This book is a result of my involvement with MATLAB teaching, both informal and in the class room, over the last several years.

This book is intended to get you started quickly. After an hour or two of getting started you can use the book as a reference. There are many examples, which you can modify for your own use. The coverage of topics is based on my experience of what is most useful, and what I wish I could have found in a book when I was learning MATLAB. If you find the book informative and useful, it is my pleasure to be of service to you. If you find it frustrating, please share your frustrations with me so that I can try to improve future editions.

The current edition has been updated for MATLAB 6. This update required checking each command and function given in this book as examples, and changing them if required. Several new features have been added that are new in MATLAB 6. New versions of software packages usually add features that their experienced users ask for. As a result, the packages and their

manuals get bigger and bigger, and more intimidating to a new user. I have
tried hard to protect the interests of a new user in this book. It has been
a struggle to keep this book lean and thin, friendly to beginners, and yet
add more features and applications. In response to emails I have received
from several readers across the globe, I have added more exercises in this
edition. I have also added substantial material in Chapter 3 (Interactive
Computation) and Chapter 5 (Applications).

Acknowledgments

I was helped through the development of this book by the encouragement,
criticism, editing, typing, and test-learning of many people, especially at
Cornell University and the Indian Institute of Science. I thank all students
who have used this book in its past forms and provided constructive criticism.
I have also been fortunate to receive feedback by email, sometimes quite
flattering, from several readers all over the world. I greatly appreciate your
words of encouragement.

I wish to thank Chris Wohlever, Mike Coleman, Richard Rand, David
Caughey, Yogendra Simha, Vijay Arakeri, Greg Couillard, Christopher D.
Hall, James R. Wohlever, John T. Demel, Jeffrey L. Cipolla, John C. Polk-
ing, Thomas Vincent, John Gibson, Sai Jagan Mohan, Konda Reddy, Sesha
Sai, Yair Hollander, Les Axelrod, Ravi Bhusan Singh Pandaw, Gujjarappa,
Manjula, The MathWorks Inc., and Cranes Software International for the
help and support they have extended to me in the development of this book.
In addition, I must acknowledge the help of three special people. Andy Ruina
has been an integral part of the development of this book all along. In fact,
he has written most of Chapter 8, the introduction to the Symbolic Math
Toolbox. That apart, his criticisms and suggestions have influenced every
page of this book. Shishir Kumar has checked all commands and programs
for compatibility with MATLAB 6, and has added several examples. My editor Peter Gordon at Oxford University Press has always been supportive
and kept his word on keeping the price of the book low.

Lastly, I thank my wife, Kalpana, for being incredibly supportive throughout. The first edition of this book came out in 1995, just after our daughter,
Manisha, was born. She learned to pronounce MATLAB at the age of two.
Now that she has graduated to recognizing the MATLAB prompt and doing
simple integer calculations in MATLAB, a new batch of absolute beginners
has arrived — twin boys Manas and Mayank. Despite their arrival, if this
edition of the book is in your hands, it is because of my wife who provided
me with the time to work on the book by shouldering more than her fair
share of family responsibilities.

Thank you all.

Bangalore   Rudra Pratap

1. *Introduction*

1.1   What Is MATLAB?

MATLAB™ is a software package for high-performance numerical computation and visualization. It provides an interactive environment with hundreds of built-in functions for technical computation, graphics, and animation. Best of all, it also provides easy extensibility with its own high-level programming language. The name MATLAB stands for MATrix LABoratory.

The diagram in Fig. 1.1 shows the main features and capabilities of MATLAB. MATLAB's built-in functions provide excellent tools for linear algebra computations, data analysis, signal processing, optimization, numerical solution of ordinary differential equations (ODEs), quadrature, and many other types of scientific computations. Most of these functions use state-of-the art algorithms. There are numerous functions for 2-D and 3-D graphics as well as for animation. Also, for those who cannot do without their Fortran or C codes, MATLAB even provides an external interface to run those programs from within MATLAB. The user, however, is not limited to the built-in functions; he can write his own functions in the MATLAB language. Once written, these functions behave just like the built-in functions. MATLAB's language is very easy to learn and to use.

There are also several *optional* 'Toolboxes' available from the developers of MATLAB. These Toolboxes are collections of functions written for special applications such as Symbolic Computation, Image Processing, Statistics, Control System Design, Neural Networks, etc.

The basic building block of MATLAB is the matrix. The fundamental data-type is the *array*. Vectors, scalars, real matrices and complex matrices are all automatically handled as special cases of the basic data-type. What is more, you almost never have to declare the dimensions of a matrix. MATLAB simply loves matrices and matrix operations. The built-in functions are

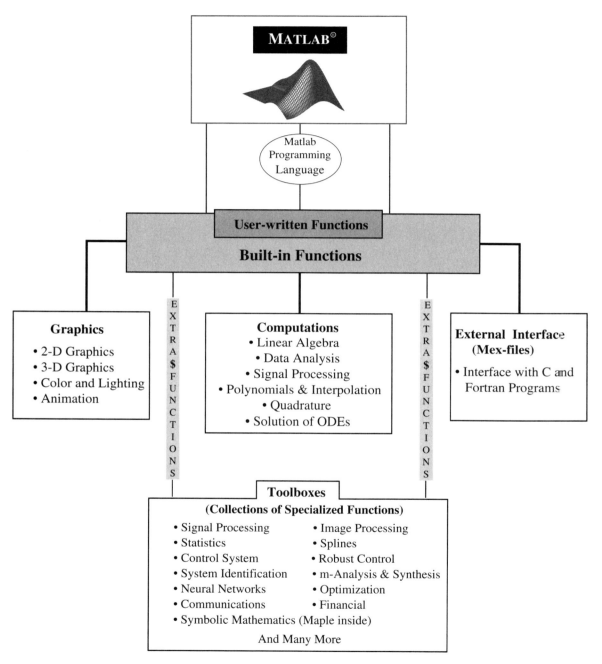

Figure 1.1: A schematic diagram of MATLAB's main features.

optimized for vector operations. Consequently, *vectorized*[1]. commands or codes run much faster in MATLAB.

1.2   Does MATLAB Do Symbolic Calculations?

(MATLAB vs Mathematica, Maple, and Macsyma)

If you are new to MATLAB, you are likely to ask this question. The first thing to realize is that MATLAB is primarily a numerical computation package, although with the Symbolic Toolbox (standard with the Student Edition of MATLAB. See Section 8.1 on page 217 for an introduction) it can do symbolic algebra [2]. Mathematica, Maple, and Macsyma are primarily symbolic algebra packages. Of course, they do *numerical* computations too. In fact, if you know any of these packages *really* well, you can do almost every calculation that MATLAB does using that software. So why learn MATLAB? Well, MATLAB's ease of use is its best feature. Also, it has a shallow learning curve (more learning with less effort) while the computer algebra systems have a steep learning curve. Since MATLAB was primarily designed to do numerical calculations and computer algebra systems were not, MATLAB is often much faster at these calculations—often as fast as C or Fortran. There are other packages, such as Xmath, that are also closer in aim and scope but seem to be popular with people in some specialized application areas. The bottom line is, in numerical computations, especially those that utilize vectors and matrices, MATLAB beats everyone hands down in terms of ease of use, availability of built-in functions, ease of programming, and speed. The proof is in the phenomenal growth of MATLAB users around the world in just a few years. There are more than 2000 universities and thousands of companies listed as registered users. MATLAB's popularity today has forced such powerful packages as Mathematica and Macsyma to provide extensions for files in MATLAB's format!

---

[1]Vectorization refers to a manner of computation in which an operation is performed simultaneously on a list of numbers (a vector) rather than sequentially on each member of the list. For example, let θ be a list of 100 numbers. Then $y = \sin(\theta)$ is a vectorized statement as opposed to $y_1 = \sin(\theta_1), y_2 = \sin(\theta_2)$, etc.

[2]Symbolic algebra means that computation is done in terms of symbols or variables rather than numbers. For example, if you type `(x+y)^2` on your computer and the computer responds by saying that the expression is equal to $x^2 + 2xy + y^2$, then your computer does symbolic algebra. Software packages that do symbolic algebra are also known as *Computer Algebra Systems*.

1.3   Will MATLAB Run on My Computer?

The most likely answer is "yes," because MATLAB supports almost every computational platform. In addition to Windows, MATLAB 6 is available for AIX, Digital UNIX, HP UX (including UX 10, UX 11), IRIX, IRIX64, Linux, and Solaris operating systems. Older versions of MATLAB are available for additional platforms such as Mac OS, and Open VMS. To find out more about product availability for your particular computer, see the MathWorks homepage on the website given below.

1.4   Where Do I Get MATLAB?

MATLAB is a product of the MathWorks, Incorporated. Contact the company for product information and ordering at the following address:

The MathWorks Inc.
3 Apple Hill Drive, Natick, MA 01760-2098
Phone: (508) 647-7000,   Fax: (508) 647-7001.
Email: info@mathworks.com
World Wide Web: http://www.mathworks.com

1.5   How Do I Use This Book?

This book is intended to serve as an introduction to MATLAB. The goal is to get started as simply as possible. MATLAB is a very powerful and sophisticated package. It takes a while to understand its real power. Unfortunately, most powerful packages tend to be somewhat intimidating to a beginner. That is why this book exists — to help you overcome the fear, get started quickly, and become productive in very little time. The most useful and easily *accessible* features of MATLAB are discussed first to make you productive and build your confidence. Several features are discussed in sufficient depth, with an invitation to explore the more advanced features on your own. All features are discussed through examples using the following conventions:

- **Typographical styles:**
  - All actual MATLAB commands or instructions are shown in `typed face`.
  - Place holders for variables or names in a command are shown in *italics*. So, a command shown as `help` *`topic`* implies that you have to type the actual name of a topic in place of *topic* in the command.
  - *Italic* text has also been used to *emphasize* a point and sometimes, to introduce a new term.

- **Actual examples:** Actual examples carried out in MATLAB are shown in gray, shaded boxes. Explanatory notes have been added within small white rectangles in the gray boxes as shown below.

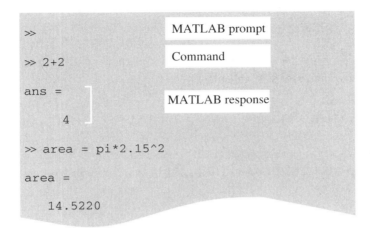

```
>>

>> 2+2

ans =

      4

>> area = pi*2.15^2

area =

    14.5220
```

| MATLAB prompt |
| Command |
| MATLAB response |

Figure 1.2: Actual examples carried out in MATLAB are shown in gray boxes throughout this book. The texts in the white boxes inside these gray boxes are explanatory notes.

These gray, boxed figures are intended to provide a parallel track for the impatient reader. If you would rather try out MATLAB right away, you are encouraged to go through these boxed examples. Most of the examples are designed so that you can (more or less) follow them without reading the entire text. All examples are system-independent. After trying out the examples, you should read the appropriate sections.

- **On-line help:** We encourage the use of on-line help. For almost all major topics, we indicate the on-line help *category* in a small box in the margin as shown here.

For on-line help type:
```
help help
```

Typing `help category` in MATLAB with the appropriate category name provides a list of functions and commands in that category. Detailed help can then be obtained for any of those commands and functions.

We discourage a passive reading of this book. The best way to learn any computer software is to try it out. We believe this, practice it, and encourage you to practice it too. So, if you are impatient, quickly read Sections 1.6.1–1.6.3, jump to the tutorials on page 19, and get going.

1.6   Basics of MATLAB

Here we discuss some basic features and commands. To begin, let us look at the general structure of the MATLAB environment.

1.6.1   MATLAB windows

On almost all systems, MATLAB works through three basic windows, which are shown in Fig. 1.3 and discussed below.

1. **Command window:** This is the main window. It is characterized by the MATLAB command prompt '≫'. When you launch the application program, MATLAB puts you in this window. All commands, including those for running user-written programs, are typed in this window at the MATLAB prompt. In MATLAB 6, this window is a part of the MATLAB window (see Fig. 1.3) that contains four other smaller windows. If you can get to the command window, we advise you to ignore the other four subwindows at this point. As software packages, such as MATLAB, become more and more powerful, their creators add more and more features to address the needs of experienced users. Unfortunately, it makes life harder for the beginners — there is more room for confusion, distraction, and intimidation. Although, we describe the other subwindows here that appear with the command window, we do not expect it to be useful to you till you get to Lesson 6 in Chapter 2.

   Launch Pad: This subwindow lists all MATLAB related applications and toolboxes that are installed on your machine. You can launch any of the listed applications by double clicking on them.

   Workspace: This subwindow lists all variables that you have generated so far and shows their type and size. You can do various things with these variables, such as plotting, by clicking on a variable and then using the right button on the mouse to select your option.

   Command History: All commands typed on the MATLAB prompt in the command window get recorded, even across multiple sessions (you worked on Monday, then on Thursday, and then on next Wednesday, and so on), in this window. You can select a command from this window with the mouse and execute it in the command window by double clicking on it. You can also select a set of commands from this window and create an *M-file* with the right click of the mouse (and selecting the appropriate option from the menu).

   Current Directory: This is where all your files from the current directory are listed. You can do file navigation here. You also have several options of what you can do with a file once you select it (with a mouse click). To see the options, click the right button of the mouse after selecting a file. You can run M-files, rename them, delete them, etc.

2. **Graphics window:** The output of all graphics commands typed in the command window are flushed to the graphics or *Figure* window, a separate gray window with (default) white background color. The user can create as many figure windows as the system memory will allow.

3. **Edit window:** This is where you write, edit, create, and save your own programs in files called '*M-files*'. You can use any text editor to carry out these tasks. On most systems, MATLAB provides its own built-in editor. However, you can use your own editor by typing the standard file-editing command that you normally use on your system. From within MATLAB , the command is typed at the MATLAB prompt following the special character '!'. The exclamation character prompts MATLAB to return the control temporarily to the local operating system, which executes the command following the '!' character. After the editing is completed, the control is returned to MATLAB. For example, on Unix systems, typing `!vi myprogram.m` at the MATLAB prompt (and hitting the return key at the end) invokes the *vi* editor on the file 'myprogram.m'. Typing `!emacs myprogram.m` invokes the *emacs* editor.

1.6.2   On-line help

- **On-line documentation:** MATLAB provides on-line help for all its built-in functions and programming language constructs. The commands `lookfor`, `help`, `helpwin`, and `helpdesk` provide on-line help. See Section 3.4 on page 69 for a description of the help facility.

- **Demo:** MATLAB has a demonstration program that shows many of its features. The program includes a tutorial introduction that is worth trying. Type `demo` at the MATLAB prompt to invoke the demonstration program, and follow the instructions on the screen.

1.6.3   Input-Output

MATLAB supports interactive computation (see Chapter 3), taking the input from the screen, and flushing the output to the screen. In addition, it can read input files and write output files (see Section 4.3.7). The following features hold for all forms of input-output:

- **Data type:**   The fundamental data-type in MATLAB is the *array*. It encompasses several distinct data objects — integers, doubles (real

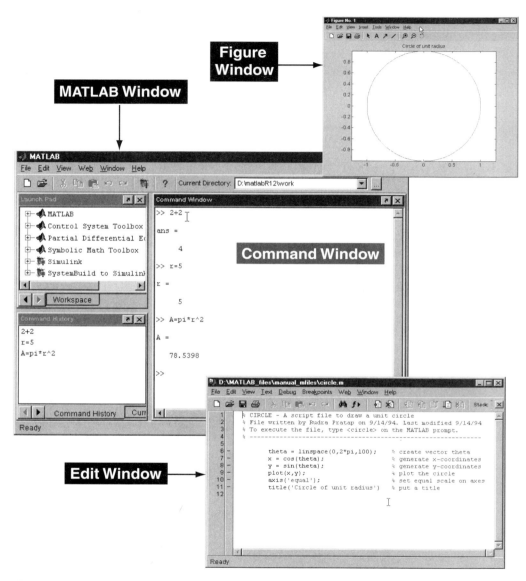

Figure 1.3: The MATLAB environment consists of a Command Window, a Figure Window, and an Edit Window. The Figure and the Editor windows appear only when invoked with the appropriate commands.

numbers), matrices, character strings, structures, and cells. [3] In most cases, however, you never have to worry about the data-type or the data object declarations. For example, there is no need to declare variables as real or complex. When a real number is entered as the value of a variable, MATLAB automatically sets the variable to be real (`double`).

- **Dimensioning:** Dimensioning is automatic in MATLAB. No dimension statements are required for vectors or arrays. You can find the dimensions of an existing matrix or a vector with the `size` and `length` (for vectors only) commands.

- **Case sensitivity:** MATLAB is case-sensitive; that is, it differentiates between the lowercase and uppercase letters. Thus `a` and `A` are different variables. Most MATLAB commands and built-in function calls are typed in lowercase letters. You can turn case sensitivity on and off with the `casesen` command. However, we do not recommend it.

- **Output display:** The output of every command is displayed on the screen unless MATLAB is directed otherwise. A semicolon at the end of a command suppresses the screen output, except for graphics and on-line help commands. The following facilities are provided for controlling the screen output :

  - **Paged output:** To direct MATLAB to show one screen of output at a time, type `more on` at the MATLAB prompt. Without it, MATLAB flushes the entire output at once, without regard to the speed at which you read.

  - **Output format:** Though computations inside MATLAB are performed using double precision, the appearance of floating point numbers on the screen is controlled by the output `format` in use. There are several different screen output formats. The following table shows the printed value of 10π in 7 different formats.

```
format short      31.4159
format short e    3.1416e+001
format long       31.41592653589793
format long e     3.141592653589793e+001
format short g    31.416
format long g     31.4159265358979
format hex        403f6a7a2955385e
format rat        3550/113
format bank       31.42
```

---

[3] *Structures* and *cells* are new data objects introduced in MATLAB 5. See Section 4.4 on page 105 for their description. MATLAB 5.x also allows users to create their own data objects and associated operations. We do not discuss this facility in this book.

The additional formats, `format compact` and `format loose`, control the spacing above and below the displayed lines, and `format +` displays a `+`, `-`, and blank for positive, negative, and zero numbers, respectively. The default is `format short`. The display format is set by typing `format` *type* on the command line (see Fig. 2.1 on page 22 for an example).

- **Command history:** MATLAB saves previously typed commands in a buffer. These commands can be recalled with the **up-arrow** key (↑). This helps in editing previous commands. You can also recall a previous command by typing the first few characters and then pressing the ↑ key. Alternatively, you can copy and paste commands from the "`Command History`" subwindow where all your commands from even previous sessions of MATLAB are recorded and listed. On most Unix systems, MATLAB's command-line editor also understands the standard *emacs* keybindings.

1.6.4　File types

MATLAB has three types of files for storing information:

M-files are standard ASCII text files, with a `.m` extension to the filename. There are two types of these files: *script files* and *function files* (see Section 4.1 and 4.2). Most programs you write in MATLAB are saved as M-files. All built-in functions in MATLAB are M-files, most of which reside on your computer in precompiled format. Some built-in functions are provided with source code in readable M-files so that they can be copied and modified.

Mat-files are binary data-files, with a `.mat` extension to the filename. Mat-files are created by MATLAB when you save data with the `save` command. The data is written in a special format that only MATLAB can read. Mat-files can be loaded into MATLAB with the `load` command (see Section 3.5 for details.).

Mex-files are MATLAB-callable Fortran and C programs, with a `.mex` extension to the filename. Use of these files requires some experience with MATLAB and a lot of patience. We do not discuss Mex-files in this introductory book.

1.6.5　Platform dependence

One of the best features of MATLAB is its platform-independence. Once you are in MATLAB, for most part, it does not matter which computer you are on. Almost all commands work the same way. The only commands that differ are the ones that necessarily depend on the local operating system, such as editing (if you do not use the built-in editor) and saving M-files. Programs written in the MATLAB language work exactly the same way on all computers. The user interface (how you interact with your computer), however, may vary a little from platform to platform.

- **Launching MATLAB :** If MATLAB is installed on your machine correctly then you can launch it by following these directions:

  On PCs: Navigate and find the MATLAB folder, locate the MATLAB program, and double-click on the program icon to launch MATLAB. If you have worked in MATLAB before and have an M-file or Mat-file that was written by MATLAB, you can also double-click on the file to launch MATLAB.

  On Unix machines: Type `matlab` on the Unix prompt and hit **return**. If MATLAB is somewhere in your *path*, it will be launched. If it is not, ask your system administrator.

- **Creating a directory and saving files:** Where should you save your files so that MATLAB can easily access them? In MATLAB 6, there is a default folder called "`work`" where MATLAB saves your files if you do not specify any other location. If you are the only user of MATLAB on the computer you are working on, this is fine. You can save all your work in this folder and access all your files easily (default set-up). If not, you have to create a separate folder for saving your work.

  Theoretically, you can create a directory/folder anywhere, save your files, and direct MATLAB to find those files. The most convenient place, however, to save all user-written files is in a directory (or folder) immediately below the directory (or folder) in which the MATLAB application program is installed (for PCs). This way all user-written files are automatically accessible to MATLAB. If you need to store the files somewhere else, you might have to specify the path to the files using the `path` command, or change the working directory of MATLAB to the desired directory with the `cd` command. We recommend the latter.

  On PCs: Create a folder inside the MATLAB folder and save your files there. If you are not allowed to write in the MATLAB folder (as may be the case in some shared facilities), then create a folder where you are allowed (perhaps on your own floppy disk), copy the file `startup.m` (if it exists[4]) from the **MATLAB/Toolbox/local** folder to your folder, and launch MATLAB by double-clicking on the `startup.m` file in your folder. This way MATLAB automatically accesses all files in your folder. You should also personalize the *Startup* file by editing it and adding a line, say, `disp('Hello Kelly, Welcome Aboard.')` You can open,

---

[4]If it does not exist, then create one using MATLAB and save it in your folder.

write, and save M-files by selecting appropriate commands from the **File** menu in MATLAB.

On Unix machines: Create a directory for your MATLAB work, save all MATLAB related files here, and launch MATLAB from this directory. To open, write, and save M-files, use a text editor such as **vi** or **emacs**. You can also use the built-in editor if you are working in MATLAB 5.2 or later versions.

- **Printing:**

    On PCs: To print the contents of the current active window (command, figure, or edit window), select **Print...** from the **File** menu and click **Print** in the dialog box. You can also print the contents of the figure window by typing `print` at the MATLAB prompt.

    On Unix machines: To print a file from inside MATLAB, type the appropriate Unix command preceded by the exclamation character (!). For example, to print the file **startup.m**, type `!lpr startup.m` on the MATLAB prompt. To print a graph that is currently in the figure window simply type `print` on the MATLAB prompt.

1.6.6   General commands you should remember

On-line help

help	lists topics on which help is available
helpwin	opens the interactive help window
helpdesk	opens the web browser based help facility
help *topic*	provides help on *topic*
lookfor *string*	lists help topics containing *string*
demo	runs the demo program

Workspace information

who	lists variables currently in the workspace
whos	lists variables currently in the workspace with their size
what	lists m-, mat-, and mex-files on the disk
clear	clears the workspace, all variables are removed
clear x y z	clears only variables x, y and z
clear all	clears all variables and functions from workspace
mlock *fun*	locks function *fun* so that clear cannot remove it
munlock *fun*	unlocks function *fun* so that clear can remove it
clc	clears command window, command history is lost
home	same as clc
clf	clears figure window

Directory information

pwd	shows the current working directory
cd	changes the current working directory
dir	lists contents of the current directory
ls	lists contents of the current directory, same as dir
path	gets or sets MATLAB search path
editpath	modifies MATLAB search path
copyfile	copies a file
mkdir	creates a directory

General information

computer	tells you the computer type you are using
clock	gives you wall clock time and date as a vector
date	tells you the date as a string
more	controls the paged output according to the screen size
ver	gives the license and the version information about MATLAB installed on your computer
bench	benchmarks your computer on running MATLAB compared to other computers

Termination

`^c` (Control-c)	local abort, kills the current command execution
`quit`	quits MATLAB
`exit`	same as `quit`

1.7   Visit This Again

We would like to point out a few things that vex the MATLAB beginners, perhaps, the most. Although, many of these things would probably not make sense to you right now, they are here, and you can come back to them whenever they seem relevant.

In the past, file navigation in MATLAB has caused considerable problem to users, especially the beginners. We have had numerous complaints from students about not being able to make MATLAB find their file, get MATLAB to work from their directory, get MATLAB to find and execute the currently edited file, etc. Fortunately, MATLAB 6 has incorporated some new features that mitigate this problem immensely. The "Current Directory" is shown just above the **Command Window** with the option of changing the current directory with just a click of the mouse. In addition, there is a **Current Directory** subwindow to the left of the **Command Window** that lists files in the current directory, gives you options of opening, loading (a .mat file), executing (a .m file), editing, etc., with the click of the right button on the mouse. You can also change the directory there or add a particular directory to the MATLAB path so that MATLAB has access to all the files in that directory automatically.

If you do not save all your MATLAB files in the default **Work** directory or folder, you need to be aware of the following issues.

1. **Not being in the right directory:** You may write and save many MATLAB programs (M-files) but MATLAB does not seem to find them. If your files are not in the current working directory, MATLAB cannot access them. Find which directory you are currently in by looking at the small **Current Directory** window in the toolbar or by querying MATLAB with the command `pwd`. Use `dir` or `ls` at the command prompt to see if MATLAB lists your files or click on the **Current Directory** tab to the left of the command window to see the listing of files in that subwindow. If you do not see your files, guide MATLAB to get to the directory where your files are. Use `cd` or `path`; `cd` is easier to use but applies only to the current session. With `path` command, you can save the path to your directory and have MATLAB automatically access your directory every time you use it. Use the on-line help to see how to set the path. Also, see Lesson-6 in the tutorials (Chapter 2).

2. **Not saving files in the correct directory:** When you edit a file in the MATLAB Editor/Debugger Window, and save it, it does not automatically mean that MATLAB Command Window has access to the directory you saved your file in. So, after saving the file, when you try to execute it and MATLAB does not find your file, go to the item (1.) above and set things right.

3. **Not overwriting an existing file while editing:** You run your program by executing your M-file, do not like the result, edit the file, and run it again; but MATLAB gives the same answer! The previously *parsed* (compiled) file is executing, MATLAB does not know about your changes. This can happen due to various reasons. Simple cure is, clear the workspace with `clear all` and execute your file.

There are various other little things that cause trouble from time to time. We point them out throughout the book wherever they raise their head.

2. *Tutorial Lessons*

The following lessons are designed to get you started quickly in MATLAB. Each lesson should take about 10–15 minutes. The lessons are intended to make you familiar with the basic facilities of MATLAB. We urge you also to do the exercises given at the end of each lesson. This will take more time, but it will teach you quite a few things. If you get stuck in the exercises, simply turn the page; answers are on the back. Most answers consist of correct commands to do the exercises. But there are several correct ways of doing the problems. So, your commands might look different than those given.

Before You Start

You need some information about the computer you are going to work on. In particular, find out:

- How to switch on the computer and get it started.
- How to log on and log off.
- Where MATLAB is installed on the computer.
- How to access MATLAB.
- Where you can write and save files—hard drive or a floppy disk.
- If there is a printer attached to the computer.

If you are working on your own computer, you will most likely know the answer to these questions. If you are working on a computer in a public facility, the system manager can help you. If you are in a class that requires working on MATLAB, your professor or TA can provide answers. In public facilities, sometimes the best thing to do is to spot a friendly person working there and ask these questions politely. People are usually nice!

If you have not read the introduction (Chapter 1), we recommend that you at least read Sections 1.6.1–1.6.3 and glance through the rest of Section 1.6 before trying the tutorials.

Here are the lessons in a nutshell:

Lesson-1: Launch MATLAB, do some simple calculations, and quit.
Key features: Learn to add, multiply, and exponentiate numbers, use trig functions, and control screen output with `format`.

Lesson-2: Create and work with arrays, vectors in particular.
Key features: Learn to create, add, and multiply vectors, use `sin` and `sqrt` functions with vector arguments, and use `linspace` to create a vector.

Lesson-3: Plot simple graphs.
Key features: Learn to plot, label, and print out a circle.

Lesson-4: Write and execute a *script file.*
Key features: Learn to write, save, and execute a script file that plots a unit circle.

Lesson-5: Write and execute a *function file.*
Key features: Learn to write, save, and execute a function file that plots a circle of any specified radius.

Lesson-6: Learn about file and directory navigation.
Key features: Learn several ways of checking your current directory, changing working directory, and setting MATLAB path.

2.1 Lesson 1: A Minimum MATLAB Session

Goal: To learn how to log on, invoke MATLAB, do a few trivial calculations, quit MATLAB, and log off.

Time Estimates:
Lesson: 10 minutes
Exercises: 30 minutes

What you are going to learn:

- How to do simple arithmetic calculations. The arithmetic operators are:

+	addition,
−	subtraction,
*	multiplication,
/	division, and
^	exponentiation.

- How to assign values to variables.
- How to suppress screen output.
- How to control the appearance of floating point numbers on the screen.
- How to quit MATLAB.

The MATLAB commands/operators used are

```
+, -, *, /, ^, ;
sin, cos, log
format
quit
```

In addition, if you do the exercises, you will learn more about arithmetic operations, exponentiation and logarithms, trigonometric functions, and complex numbers.

Method: Log on and launch MATLAB. Once the MATLAB command window is on the screen, you are ready to carry out the first lesson. Some commands and their output are shown below. Go ahead and reproduce the results.

```
>> 2+2

ans =

      4

>> x = 2+2

x =

      4

>> y = 2^2 + log(pi)*sin(x);

>> y

y =

    3.1337

>> theta = acos(-1)

theta =

    3.1416

>> format short e
>> theta

theta =

  3.1416e+000

>> format long
>> theta

theta =

  3.14159265358979

>> quit
```

Enter 2+2 and hit return/enter key. Note that the result of an un-assigned expression is saved in the default variable 'ans'.

You can also assign the value of an expression to a variable.

A semicolon at the end suppresses screen output. MATLAB remembers y, though. You can recall the value of y by simply typing y.

MATLAB knows trigonometry. Here is arccosine of -1.

The floating point output display is controlled by the format command. Here are two examples. More info on this later.

Quit MATLAB. You can also quit by selecting quit from the file menu on Macs and PCs.

Figure 2.1: Lesson-1: Some simple calculations in MATLAB.

EXERCISES

1. **Arithmetic operations:** Compute the following quantities:
   - $\frac{2^5}{2^5-1}$ and compare with $(1 - \frac{1}{2^5})^{-1}$.
   - $3\frac{\sqrt{5}-1}{(\sqrt{5}+1)^2} - 1$. The square root $\sqrt{x}$ can be calculated with the command `sqrt(x)` or `x^0.5`.
   - Area $= \pi r^2$ with $r = \pi^{\frac{1}{3}} - 1$. ($\pi$ is pi in MATLAB.)

2. **Exponential and logarithms:** The mathematical quantities e^x, $\ln x$, and $\log x$ are calculated with `exp(x)`, `log(x)`, and `log10(x)`, respectively. Calculate the following quantities:
   - e^3, $\ln(e^3)$, $\log_{10}(e^3)$, and $\log_{10}(10^5)$.
   - $e^{\pi\sqrt{163}}$.
   - Solve $3^x = 17$ for x and check the result. (The solution is $x = \frac{\ln 17}{\ln 3}$. You can verify the result by direct substitution.)

3. **Trigonometry:** The basic MATLAB trig functions are `sin`, `cos`, `tan`, `cot`, `sec`, and `csc`. The inverses, e.g., arcsin, arctan, etc., are calculated with `asin`, `atan`, etc. The same is true for hyperbolic functions. The inverse function `atan2` takes 2 arguments, `y` and `x`, and gives the four-quadrant inverse tangent. The argument of these functions must be in radians.
   Calculate the following quantities:
   - $\sin\frac{\pi}{6}$, $\cos\pi$, and $\tan\frac{\pi}{2}$.
   - $\sin^2\frac{\pi}{6} + \cos^2\frac{\pi}{6}$. (Typing `sin^2(x)` for $\sin^2 x$ will produce an error).
   - $y = \cosh^2 x - \sinh^2 x$, with $x = 32\pi$.

4. **Complex numbers:** MATLAB recognizes the letters `i` and `j` as the imaginary number $\sqrt{-1}$. A complex number $2 + 5i$ may be input as `2+5i` or `2+5*i` in MATLAB . The former case is always interpreted as a complex number whereas the latter case is taken as complex only if `i` has not been assigned any local value. The same is true for `j`. This kind of context dependence, for better or worse, pervades MATLAB. Compute the following quantities.
   - $\frac{1+3i}{1-3i}$. Can you check the result by hand calculation?
   - $e^{i\frac{\pi}{4}}$. Check the Euler's Formula $e^{ix} = \cos x + i\sin x$ by computing the right hand side too, i. e., compute $\cos(\pi/4) + i\sin(\pi/4)$.
   - Execute the commands `exp(pi/2*i)` and `exp(pi/2i)`. Can you explain the difference between the two results?

Answers to Exercises

1. **Command**   **Result**
   ```
   2^5/(2^5-1)                                                          1.0323
   3*(sqrt(5)-1)/(sqrt(5)+1)^2 - 1                                      -0.6459
   area=pi*(pi^(1/3)-1)^2                                               0.6781
   ```

2. **Command**   **Result**
   ```
   exp(3)                                                               20.0855
   log(exp(3))                                                          3.0000
   log10(exp(3))                                                        1.3029
   log10(10^5)                                                          5.0000
   exp(pi*sqrt(163))                                                    2.6254e+017
   x=log(17)/log(3)                                                     2.5789
   ```

3. **Command**   **Result**
   ```
   sin(pi/6)                                                            0.5000
   cos(pi)                                                              -1.0000
   tan(pi/2)                                                            1.6332e+016
   (sin(pi/6))^2+(cos(pi/6))^2                                          1
   x=32*pi; y=(cosh(x))^2-(sinh(x))^2                                   0
   ```

4. **Command**   **Result**
   ```
   (1+3i)/(1-3i)                                                        -0.8000 + 0.6000i
   exp(i*pi/4)                                                          0.7071 + 0.7071i
   exp(pi/2*i)                                                          0.0000 + 1.0000i
   exp(pi/2i)                                                           0.0000 - 1.0000i
   ```

   Note that

   `exp(pi/2*i)` $= e^{\frac{\pi}{2}i} = \cos(\frac{\pi}{2}) + i\,\sin(\frac{\pi}{2}) = i$

   `exp(pi/2i)` $= e^{\frac{\pi}{2i}} = e^{-\frac{\pi}{2}i} = \cos(\frac{\pi}{2}) - i\,\sin(\frac{\pi}{2}) = -i$

2.2   Lesson 2: Creating and Working with Arrays of Numbers

Goal: To learn how to create arrays and vectors, and how to perform arithmetic and trigonometric operations on them.

An *array* is a list of numbers or expressions arranged in horizontal rows and vertical columns. When an array has only one row or column, it is called a *vector*. An array with m rows and n columns is a called a *matrix* of size $m \times n$. See Section 3.1 for more information.

Time Estimates:
Lesson: 15 minutes
Exercises: 45 minutes

What you are going to learn:

- How to create row and column vectors.
- How to create a vector of n numbers linearly (equally) spaced between two given numbers a and b.
- How to do simple arithmetic operations on vectors.
- How to do *array operations*:

.*	term by term multiplication,
./	term by term division, and
.^	term by term exponentiation.

- How to use trigonometric functions with array arguments.
- How to use elementary math functions such as square root, exponentials, and logarithms, with array arguments.

This lesson deals primarily with one-dimensional arrays, i.e., vectors. One of the exercises introduces you to two-dimensional arrays, i.e., matrices. There are many mathematical concepts associated with vectors and matrices that we do not mention here. If you have some background in linear algebra, you will find that MATLAB is set up to do almost any matrix computation (e.g., inverse, determinant, rank, etc.).

Method: You already know how to launch MATLAB. So go ahead and try the commands shown on the next page. Once again, you are going to reproduce the results shown.

```
>> x = [1 2 3]                          x is a row vector with 3 elements.

x =
       1      2      3

>> y = [2; 1; 5]
                                        y is a column vector with 3
y =                                     elements.
       2
       1
       5

>> z = [2 1 0];
>> a = x + z                            You can add (or subtract) two
                                        vectors of the same size.

a =
       3      3      3

>> b = x + y                            But you cannot add (or subtract)
                                        a row vector to a column vector.
??? Error using ==> +
Matrix dimensions must agree.

>> a = x.*z                             You can multiply (or divide) the
                                        elements of two same-sized
a =                                     vectors term by term with the
       2      2      0                  array operator .* (or ./).

>> b = 2*a                              But multiplying a vector with a
                                        scalar does not need any special
b =                                     operation (no dot before the *).
       4      4      0

>> x = linspace(0,10,5)                 Create a vector x with 5 elements
                                        linearly spaced between 0 and 10.
x =
       0   2.5000   5.0000   7.5000  10.0000

>> y = sin(x);                          Trigonometric functions sin, cos.
                                        etc., as well as elementary math
>> z = sqrt(x).*y                       functions sqrt, exp, log, etc.,
                                        operate on vectors term by term.
z =
       0   0.9463  -2.1442   2.5688  -1.7203
```

Figure 2.2: Lesson-2: Some simple calculations with vectors.

EXERCISES

1. **Equation of a straight line:** The equation of a straight line is $y = mx + c$ where m and c are constants. Compute the y-coordinates of a line with slope $m = 0.5$ and the intercept $c = -2$ at the following x-coordinates:

$$x = 0, \quad 1.5, \quad 3, \quad 4, \quad 5, \quad 7, \quad 9, \text{ and } 10.$$

   [Note: Your command should not involve any array operators since your calculation involves multiplication of a vector with a scalar m and then addition of another scalar c.]

2. **Multiply, divide, and exponentiate vectors:** Create a vector t with 10 elements: 1, 2, 3, ..., 10. Now compute the following quantities:

   - $x = t \sin(t)$.
   - $y = \frac{t-1}{t+1}$.
   - $z = \frac{\sin(t^2)}{t^2}$.

3. **Points on a circle:** All points with coordinates $x = r \cos \theta$ and $y = r \sin \theta$, where r is a constant, lie on a circle with radius r, i.e., they satisfy the equation $x^2 + y^2 = r^2$. Create a column vector for θ with the values 0, $\pi/4$, $\pi/2$, $3\pi/4$, π, and $5\pi/4$.

   Take $r = 2$ and compute the column vectors x and y. Now check that x and y indeed satisfy the equation of circle, by computing the radius $r = \sqrt{(x^2 + y^2)}$. [To calculate r you will need the array operator `.^` for squaring x and y. Of course, you could compute x^2 by `x.*x` also.]

4. **The geometric series:** This is funky! You know how to compute x^n element-by-element for a vector x and a scalar exponent n. How about computing n^x, and what does it mean? The result is again a vector with elements n^{x_1}, n^{x_2}, n^{x_3} etc.

   The sum of a geometric series $1 + r + r^2 + r^3 + \ldots + r^n$ approaches the limit $\frac{1}{1-r}$ for $r < 1$ as $n \to \infty$. Create a vector n of 11 elements from 0 to 10. Take $r = 0.5$ and create another vector $x = [r^0 \quad r^1 \quad r^2 \quad \ldots \quad r^n]$ with the command `x = r.^n` . Now take the sum of this vector with the command `s = sum(x)` (s is the sum of the actual series). Calculate the limit $\frac{1}{1-r}$ and compare the computed sum s. Repeat the procedure taking n from 0 to 50 and then from 0 to 100.

5. **Matrices and vectors:** Go to Fig. 3.1 on page 51 and reproduce the results. Now create a vector and a matrix with the following commands: `v = 0:0.2:12;` and `M = [sin(v); cos(v)];` (see Section 3.1.4 on page 55 for use of ':' in creating vectors). Find the sizes of `v` and `M` using the **size** command. Extract the first 10 elements of each row of the matrix, and display them as column vectors.

Answers to Exercises

Commands to solve each problem are given below.

1. `x=[0 1.5 3 4 5 7 9 10];`
   `y = 0.5*x-2`
   Ans. $y = [-2.0000 \ -1.2500 \ -0.5000 \ 0 \ 0.5000 \ 1.5000 \ 2.5000 \ 3.0000].$

2. `t=1:10;`
   `x = t.*sin(t)`
   `y = (t-1)./(t+1)`
   `z = sin(t.^2)./(t.^2)`

3. `theta = [0;pi/4;pi/2;3*pi/4;pi;5*pi/4]`
   `r = 2;`
   `x=r*cos(theta); y=r*sin(theta);`
   `x.^2 + y.^2`

4. `n = 0:10;`
   `r = 0.5; x = r.^n;`
   `s1 = sum(x)`
   `n=0:50; x=r.^n; s2=sum(x)`
   `n=0:100; x=r.^n; s3=sum(x)`
   [*Ans.* $s1 = 1.9990$, $s2 = 2.0000$, and $s3 = 2$]

5. `v=0:0.2:12;`
   `M=[sin(v); cos(v)];`
   `size(v)`
   `size(M)`
   `M(:,1:10)'`
   [*Ans.* v is 1×61 and M is 2×61.
   The last command `M(:,1:10)'` picks out the first 10 elements from each row of M and transposes to give a 10×2 matrix.]

2.3   Lesson 3: Creating and Printing Simple Plots

Goal: To learn how to make a simple 2-D plot in MATLAB and print it out.

Time Estimates:
Lesson: 10 minutes
Exercises: 40 minutes

What you are going to learn:

- How to generate x and y coordinates of 100 equidistant points on a unit circle.
- How to plot x vs y and thus create the circle.
- How to set the scale of the x and y axes to be the same, so that the circle looks like a circle and not an ellipse.
- How to label the axes with text strings.
- How to title the graph with a text string.
- How to get a hardcopy of the graph.

The MATLAB commands used are

plot	creates a 2-D line plot
axis	changes the aspect ratio of x and y axes
xlabel	annotates the x-axis
ylabel	annotates the y-axis
title	puts a title on the plot, and
print	prints a hardcopy of the plot.

This lesson teaches you the most basic graphics commands. The exercises take you through various types of plots, overlay plots, and more involved graphics.

Method: You are going to draw a circle of unit radius. To do this, first generate the data (x- and y-coordinates of, say, 100 points on the circle), then plot the data, and finally print the graph. For generating data, use the parametric equation of a unit circle:

$$x = \cos\theta, \quad y = \sin\theta, \quad 0 \leq \theta \leq 2\pi.$$

In the sample session shown here, only the commands are shown. You should see the output on your screen.

`>> theta = linspace(0,2*pi,100);`	Create a linearly spaced 100 elements long vector `theta`.
`>> x = cos(theta);`	
`>> y = sin(theta);`	Calculate x and y coordinates.
`>> plot(x,y)`	Plot x vs. y. (see Section 6.1)
`>> axis('equal');`	Set the length scales of the two axes to be the same.
`>> xlabel('x')`	Label the x-axis with x.
`>> ylabel('y')`	Label the y-axis with y.
`>> title('Circle of unit radius')`	Put a title on the plot.
`>> print`	Print on the default printer.

Figure 2.3: Lesson-3: Plotting and printing a simple graph.

Comments:

- After you enter the command `plot(x,y)`, you should see an ellipse in the Figure Window. MATLAB draws an ellipse rather than a circle because of its default rectangular axes. The next command `axis('equal')`, directs MATLAB to use the same scale on both axes, so that a circle appears as a circle. You can also use `axis('square')` to override the default rectangular axes.

- The arguments of the `axis`, `xlabel`, `ylabel`, and `title` commands are text strings. Text strings are entered within single right-quote (') characters. For more information on text strings, see Section 3.2.6 on page 63.

- The `print` command sends the current plot to the printer connected to your computer.

EXERCISES

1. **A simple sine plot:** Plot $y = \sin x$, $\quad 0 \le x \le 2\pi$, taking 100 linearly spaced points in the given interval. Label the axes and put 'Plot created by *yourname*' in the title.

2. **Line-styles:** Make the same plot as above, but rather than displaying the graph as a curve, show the unconnected data points. To display the data points with small circles, use `plot(x,y,'o')`. [Hint: You may peep into Section 6.1 on page 159 if you wish.] Now combine the two plots with the command `plot(x,y,x,y,'o')` to show the line through the data points as well as the distinct data points.

3. **An exponentially decaying sine plot:** Plot $y = e^{-0.4x} \sin x$, $\quad 0 \le x \le 4\pi$, taking 10, 50, and 100 points in the interval. [Be careful about computing y. You need array multiplication between `exp(-0.4*x)` and `sin(x)`. See Section 3.2.1 on page 58 for more discussion on array operations.]

4. **Space curve:** Use the command `plot3(x,y,z)` to plot the circular helix $x(t) = \sin t$, $\; y(t) = \cos t$, $\; z(t) = t$, $\quad 0 \le t \le 20$.

5. **On-line help:** Type `help plot` on the MATLAB prompt and hit return. If too much text flashes by the screen, type `more on`, hit return, and then type `help plot` again. This should give you paged screen output. Read through the on-line help. To move to the next page of the screen output, simply press the spacebar.

6. **Log scale plots:** The plot commands `semilogx`, `semilogy`, and `loglog`, plot the x-values, the y-values, and both x- and y-values on a $\log_{10}$ scale, respectively.

   Create a vector `x = 0:10:1000`. Plot x vs. x^3 using the three log scale plot commands. [Hint: First, compute `y=x.^3` and then, use `semilogx(x,y)` etc.]

7. **Overlay plots:** Plot $y = \cos x$ and $z = 1 - \frac{x^2}{2} + \frac{x^4}{24}$ for $0 \le x \le \pi$ on the same plot. You might like to read Section 6.1.5 on page 163 to learn how to plot multiple curves on the same graph. [Hint: You can use `plot(x,y,x,z,'--')` or you can plot the first curve, use the `hold on` command, and then plot the second curve on top of the first one.]

8. **Fancy plots:** Go to Section 6.1.6 on page 167 and look at the examples of specialized 2D plots on pages 168–171. Reproduce any of the plots you like.

9. **A very difficult plot:** Use your knowledge of *splines* and *interpolation* to draw a lizard (just kidding).

Answers to Exercises

Commands required to solve the problems are shown below.

1. ```
   x=linspace(0,2*pi,100);
   plot(x,sin(x))
   xlabel('x'), ylabel('sin(x)')
   title('Plot created by Rudra Pratap')
   ```

2. ```
   plot(x,sin(x),x,sin(x),'o')
   xlabel('x'), ylabel('sin(x)')
   ```

3. ```
   x=linspace(0,4*pi,10);   % with 10 points
   y=exp(-.4*x).*sin(x);
   plot(x,y)
   x=linspace(0,4*pi,50);   % with 50 points
   y=exp(-.4*x).*sin(x);
   plot(x,y)
   x=linspace(0,4*pi,100); % with 100 points
   y=exp(-.4*x).*sin(x);
   plot(x,y)
   ```

4. ```
   t=linspace(0,20,100);
   plot3(sin(t),cos(t),t)
   ```

5. You should not be looking for answer here.

6. ```
   x=0:10:1000;
   y=x.^3;
   semilogx(x,y)
   semilogy(x,y)
   loglog(x,y)
   ```

7. ```
   x=linspace(0,pi,100);
   y=cos(x); z=1-x.^2/2+x.^4/24;
   plot(x,y,x,z)
   plot(x,y,x,z,'--')
   legend('cos(x)','z')      % try this legend command
   ```

   [For fun: If the last command **legend** does produce a legend on your plot, click and hold your mouse on the legend and see if you can move it to a location of your liking. See page 161 for more information on **legend**.]

2.4   Lesson 4: Creating, Saving, and Executing a Script File

Goal: To learn how to create Script files and execute them in MATLAB.

A *Script File* is a user-created file with a sequence of MATLAB commands in it. The file must be saved with a '.m' extension to its name, thereby, making it an *M-file*. A script file is executed by typing its name (without the '.m' extension') at the command prompt. For more information, see Section 4.1 on page 85.

Time Estimates:
Lesson: 20 minutes
Exercises: 30 minutes

What you are going to learn:

- How to create, write, and save a script file.
- How to execute the script file in MATLAB.

Unfortunately, creating, editing, and saving files are somewhat system dependent tasks. The commands needed to accomplish these tasks depends on the operating system and the text editor you use. It is not possible to provide an introduction to these topics here. So, we assume that

- You know how to use a text editor on your Unix system (for example, **vi** or **emacs**), or that you're using the built-in MATLAB editor on a Mac or a PC.
- You know how to open, edit, and save a file.
- You know which directory your file is saved in.

Method: Write a script file to draw the unit circle of Lesson-3. You are essentially going to write the commands shown in Fig. 2.3 in a file, save it, name it, and execute it in MATLAB. Follow the directions below.

1. Create a new file:

    - **On PC's and Macs:** Select New M-File from the File menu. A new edit window should appear.
    - **On Unix workstations:** Type `!vi circle.m` or `!emacs circle.m` at the MATLAB prompt to open an edit window in **vi** or **emacs**.

2. Type the following lines into this file. Lines starting with a % sign are interpreted as comment lines by MATLAB and are ignored.

```
% CIRCLE - A script file to draw a unit circle
% File written by Rudra Pratap. Last modified 6/28/98
% ------------------------
theta = linspace(0,2*pi,100);      % create vector theta
x = cos(theta);                    % generate x-coordinates
y = sin(theta);                    % generate y-coordinates
plot(x,y);                         % plot the circle
axis('equal');                     % set equal scale on axes
title('Circle of unit radius')     % put a title
```

3. Write and save the file under the name **circle.m**:

   - **On PC's:** Select Save As... from the File menu. A dialog box should appear. Type the name of the document as `circle.m`. Make sure the file is being saved in the folder you want it to be in (the current working folder/directory of MATLAB). Click Save to save the file.

   - **On Unix workstations:** You are on your own to write and save the file using the editor of your choice. After writing the file, quit the editor to get back to MATLAB.

4. Now get back to MATLAB and type the following commands in the command window to execute the script file.

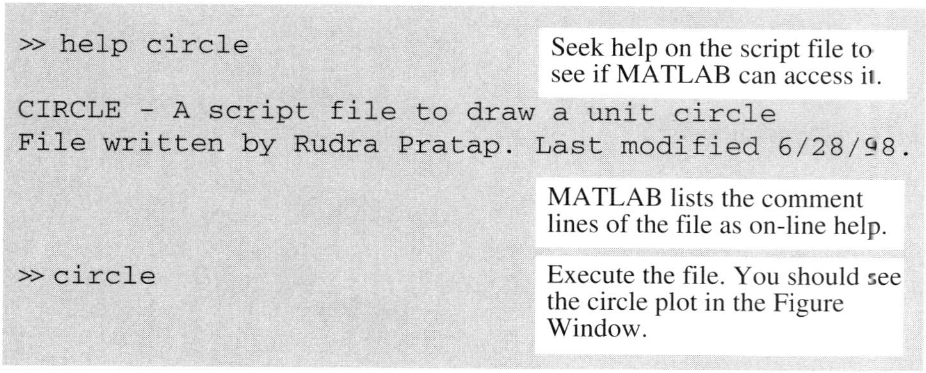

```
>> help circle                          Seek help on the script file to
                                        see if MATLAB can access it.

CIRCLE - A script file to draw a unit circle
File written by Rudra Pratap. Last modified 6/28/98.

                                        MATLAB lists the comment
                                        lines of the file as on-line help.

>> circle                               Execute the file. You should see
                                        the circle plot in the Figure
                                        Window.
```

Figure 2.4: Lesson-4: Executing a script file.

EXERCISES

1. **Show the center of the circle:** Modify the script file `circle` to show the center of the circle on the plot, too. Show the center point with a '+'. (Hint: See Exercises 2 and 7 of Lesson 3.)

2. **Change the radius of the circle:** Modify the script file `circle.m` to draw a circle of arbitrary radius r as follows:

   - Include the following command in the script file before the first executable line (`theta = ...`) to ask the user to input (r) on the screen:

     `r = input('Enter the radius of the circle:  ')`
   - Modify the x and y coordinate calculations appropriately.
   - Save and execute the file. When asked, enter a value for the radius and press return.

3. **Variables in the workspace:** All variables created by a script file are left in the global workspace. You can get information about them and access them, too:

   - Type `who` to see the variables present in your workspace. You should see the variables `r, theta, x` and `y` in the list.
   - Type `whos` to get more information about the variables and the workspace.
   - Type `[theta' x' y']` to see the values of θ, x and y listed as three columns. All three variables are row vectors. Typing a single right quote (`'` on the keyboard) after their names transposes them and makes them column vectors.

4. **Contents of the file:** You can see the contents of an M-file without opening the file with an editor. The contents are displayed by the `type` command. To see the contents of `circle.m`, type `type circle.m`.

5. **H1 line:** The first commented line before any executable statement in a script file is called the *H1 line*. It is this line that is searched by the `lookfor` command. Since the `lookfor` command is used to look for M-files with keywords in their description, you should put keywords in H1 line of all M-files you create. Type `lookfor unit` to see what MATLAB comes up with. Does it list the script file you just created?

6. **Just for fun:** Write a script file that, when executed, greets you, displays the date and time, and curses your favorite TA or professor. [The commands you need are `disp`, `date`, `clock`, and possibly `fix`. See the on-line help on these commands before using them.]

Answers to Exercises

1. Replace the command `plot(x,y)` by the command `plot(x,y,0,0,'+')`.

2. Your changed script file should look like this:

```
% CIRCLE - A script file to draw a unit circle
% File written by Rudra Pratap on 9/14/94.
%                    Last modified 6/28/98
% ----------------------------
r = input('Enter the radius of the circle: ')
theta = linspace(0,2*pi,100);      % create vector theta
x = r*cos(theta);                  % generate x-coordinates
y = r*sin(theta);                  % generate y-coordinates
plot(x,y);                         % plot the circle
axis('equal');                     % set equal scale on axes
title('Circle of given radius r')  % put a title
```

6. Here is a script file that you may not fully understand yet. Do not worry, just copy it if you like it. See the on-line help on the commands used, e.g. `disp`, `date`, `fix`, `clock`, `int2str`.

```
% Script file to begin your day. Save it as Hi_there.m
% To execute, just type Hi_there
% File written by Rudra Pratap on 6/15/95.
%                    Last modified 6/28/98
% -----------------------------
disp('Hello R.P., How is life?')
disp(' ')                         % display a blank line
disp('Today is...')
disp(date)                        % display date
time=fix(clock);                  % get time as integers
hourstr=int2str(time(4));         % get the hour
minstr=int2str(time(5));          % get the minute
if time(5)<10                     % if minute is, say 5, then
    minstr=['0',minstr];          %- write it as 05.
end
timex = [hourstr ':' minstr];     % create the time string
disp(' ')
disp('And the time is..')
disp(timex)                       % display the time
```

2.5   Lesson 5: Creating and Executing a Function File

Goal: To learn how to write and execute a *function file*. Also, to learn the difference between a script file and a function file.

A *function file* is also an M-file, just like a script file, except it has a function definition line on the top that defines the input and output explicitly. For more information, see Section 4.2.

Time Estimates:

Lesson: 15 minutes

Exercises: 60 minutes

What you are going to learn:

- How to open and edit an existing M-file.
- How to define and execute a function file.

Method: Write a function file to draw a circle of a specified radius, with the radius as the input to the function. You can either write the function file from scratch or modify the script file of Lesson 4. We advise you to select the latter option.

1. Open the script file **circle.m**:

   - **On PC's:** Select Open M-File from the File menu. Navigate and select the file **circle.m** from the Open dialog box. Double click to open the file. The contents of the file should appear in an edit window.

   - **On Unix workstations:** Type !vi circle.m or !emacs circle.m on the MATLAB prompt to open the file in a **vi** or **emacs** window.

2. Edit the file **circle.m** from Lesson–4, 34 to look like the following.

```
function [x,y] = circlefn(r);
% CIRCLEFN - Function to draw a circle of radius r.
% File written by Rudra Pratap on 9/17/94. Last modified 7/1/98
% Call syntax:   [x,y] = circlefn(r); or just:  circlefn(r);
% Input:       r = specified radius
% Output: [x,y] = the x- and y-coordinates of data points
theta = linspace(0,2*pi,100);  % create vector theta
x = r*cos(theta);              % generate x-coordinates
y = r*sin(theta);              % generate y-coordinates
plot(x,y);                     % plot the circle
axis('equal');                 % set equal scale on axes
title(['Circle of radius r =',num2str(r)])
                               % put a title with the value of r.
```

3. Now write and save the file under the name **circlefn.m** as follows:

- **On PC's:** Select **Save As...** from the **File** menu. A dialog box should appear. Type the name of the document as `circlefn.m` (usually, MATLAB automatically writes the name of the function in the document name). Make sure the file is saved in the folder you want (the current working folder/directory of MATLAB). Click **save** to save the file.

- **On Unix workstations:** You are on your own to write and save the file using the editor of your choice. After writing the file, quit the editor to get back to MATLAB.

4. Here is a sample session that executes the function **circlefn** in three different ways. Try it out.

`>> R = 5;` `>> [x,y] = circlefn(R);`	Specify the input and execute the function with an explicit output list.
`>> [cx,cy] = circlefn(2.5);`	You can also specify the value of the input directly.
`>> circlefn(1);`	If you don't need the output, you don't have to specify it.
`>> circlefn(R^2/(R+5*sin(R)));`	Of course, the input can also be a valid MATLAB expression.

Figure 2.5: Lesson-5: Executing a function file.

Comments:

- Note that a function file (see previous page) must begin with a function definition line. To learn more about function files, refer to Section 4.2 on page 88.

- The argument of the **title** command in this function file is slightly complicated. To understand how it works see Section 3.2.6 on page 63.

EXERCISES

1. **On-line help:** Type `help function` to get on-line help on `function`. Read through the help file.

2. **Convert temperature:** Write a function that outputs a conversion-table for Celsius and Fahrenheit temperatures. The input of the function should be two numbers: T_i and T_f, specifying the lower and upper range of the table in Celsius. The output should be a two column matrix: the first column showing the temperature in Celsius from T_i to T_f in the increments of 1°C and the second column showing the corresponding temperatures in Fahrenheit. To do this, (i) create a column vector C from T_i to T_f with the command `C = [Ti:Tf]'`, (ii) calculate the corresponding numbers in Fahrenheit using the formula $[F = \frac{9}{5}C + 32]$, and (iii) make the final matrix with the command `temp = [C F];`. Note that your output will be named `temp`.

3. **Calculate factorials:** Write a function `factorial` to compute the factorial $n!$ for any integer n. The input should be the number n and the output should be $n!$. You might have to use a *for* loop or a *while* loop to do the calculation. See Section 4.3.4, page 99 for a quick description of these loops. (You can use the built-in function `prod` to calculate factorials. For example, $n!$ =`prod(1:n)`. In this exercise, however, do not use this function.)

4. **Compute the cross product:** Write a function file `crossprod` to compute the cross product of two vectors **u**, and **v**, given $\mathbf{u} = (u_1, u_2, u_3)$, $\mathbf{v} = (v_1, v_2, v_3)$, and $\mathbf{u} \times \mathbf{v} = (u_2 v_3 - u_3 v_2, u_3 v_1 - u_1 v_3, u_1 v_2 - u_2 v_1)$. Check your function by taking cross products of pairs of unit vectors: $(\mathbf{i}, \mathbf{j})$, $(\mathbf{j}, \mathbf{k})$, etc. $[\mathbf{i} = (1, 0, 0), \mathbf{j} = (0, 1, 0), \mathbf{k} = (0, 0, 1)]$. (Do not use the built-in function `cross` here.)

5. **Sum a geometric series:** Write a function to compute the sum of a geometric series $1 + r + r^2 + r^3 + \ldots + r^n$ for a given r and n. Thus the input to the function must be r and n and the output must be the sum of the series. [See Exercise 4 of Lesson 2.]

6. **Calculate the interest on your money:** The interest you get at the end of n years, at a flat annual rate of $r\%$, depends on how the interest is compounded. If the interest is added to your account k times a year, and the principal amount you invested is X_0, then at the end of n years you would have $X = X_0 \left(1 + \frac{r}{k}\right)^{kn}$ amount of money in your account. Write a function to compute the interest $(X - X_0)$ on your account for a given X, n, r, and k.

   Use the function to find the difference between the interest paid on $1000 at the rate of 6% a year at the end of 5 years if the interest is compounded (i) quarterly ($k = 4$) and (ii) daily ($k = 365$). For screen output, use `format bank`.

Answers to Exercises

Some of the commands in the following functions might be too advanced for you at this point. If so, look them up or ignore them.

2.

```
function temptable = ctof(tinitial,tfinal);
% CTOF : function to convert temperature from C to F
% call syntax:
%          temptable = ctof(tinitial,tfinal);
% ------------
C = [tinitial:tfinal]';       % create a column vector C
F = (9/5)*C + 32;             % compute corresponding F
temptable = [C F];            % make a 2 column matrix of C & F.
```

3.

```
function factn = factorial(n);
% FACTORIAL: function to compute factorial n!
% call syntax:
%          factn = factorial(n);
% ------------
factn = 1;                    % initialize. also 0! = 1.
for k = n:-1:1                % go from n to 1
    factn = factn*k;         % multiply n by n-1, n-2 etc.
end
```

Can you modify this function to check for negative input and non-integer input before it computes the factorial?

4.

```
function w = crossprod(u,v);
% CROSSPROD: function to compute w = u x v for vectors u & v.
% call syntax:
%          w = crossprod(u,v);
% ------------
if length(u)>3 | length(v)>3  % check if u OR v has > 3 elements
    error('Ask Euler. This cross product is beyond me.')
end
w = [u(2)*v(3)-u(3)*v(2);      % first element of w
     u(3)*v(1)-u(1)*v(3);      % second element of w
     u(1)*v(2)-u(2)*v(1)];     % third element of w
```

Can you modify this function to check for 2-D vectors and set the third component automatically to zero?

5.
```
function s = gseriessum(r,n);
% GSERIESSUM: function to calculate the sum of a geometric series
% The series is  1+r+r^2+r^3+....r^n (upto nth power).
% call syntax:
%          s = gseriessum(r,n);
% ------------
nvector = 0:n;                  % create a vector from 0 to n
series = r.^nvector;            % create a vector of terms in the series
s = sum(series);                % sum all elements of the vector 'series'.
```

6.
```
function [capital,interest] = compound(capital,years,rate,timescomp);
% COMPOUND: function to compute the compounded capital and the interest
% call syntax:
%          [capital,interest] = compound(capital,years,rate,timescomp);
% ------------
x0 = capital; n = years; r = rate; k = timescomp;
if r>1                        % check for common mistake
   disp('check your interest rate. For 8% enter .08, not 8.')
end
capital = x0*(1+r/k)^(k*n);
interest = capital - x0;
```

[*Ans.* (i) Quarterly: $346.85,   Daily: $349.83,   Difference: $ 2.98.]

2.6   Lesson 6: Working with Files and Directories

Goal: To learn how to navigate through MATLAB directory system and work with files in a particular location.

Time Estimates:
Lesson: 30 minutes
Exercises: 0 minutes

What you are going to learn:

- How to find your bearings in the jungle of directories.
- How to find which of your M-files are accessible.
- How to change the working directory.
- How to set MATLAB path so that it finds your favorite directories automatically.

Method: MATLAB 6 includes several menu driven features which make file navigation much easier (compared to the earlier versions). You will explore some of these features now. In addition, you will also learn commands that pretty much do the same thing from the command line. The commands that you will use are `pwd, dir, ls, cd, what`, and `path`. Let us go step by step.

Where are you? The first thing to find is which directory you are currently in. This information is available in three ways:

1. Look at the command window toolbar. There is a small window that shows the name of the current directory along with its path. For example, Fig. 2.6 shows that the current directory is `D:\matlabR12\work`. This is the default directory that MATLAB puts you in. As the path indicates, it is inside the `matlabR12` directory.

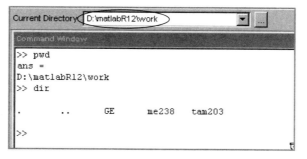

Figure 2.6: Which directory are you in?

2. You can get the same information from the command line by typing `pwd` (print working directory).

3. The current directory is also displayed in a separate subwindow to the left of the command window. If it is not visible, click on the **Current Directory** tab. This subwindow also lists the contents of the current directory.

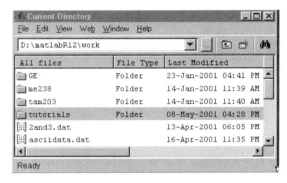

Figure 2.7: Current directory information from the MATLAB desktop.

How do you change the current directory? You can change the current directory with the `cd` *DirectoryName* command on the command line or by navigating through the browse button (the button with three dots on it) located next to the current directory peep-in window. Make sure that after you change the directory, the new directory's name appears in the current directory peep-in window.

What are the contents of the current directory? You can see the contents of the current directory in the **Current Directory** subwindow (Fig. 2.7) or by typing `dir` or `ls` on the command line. These commands list all the files and folders in the current directory. If you would much rather see only MATLAB related files (e.g., M-files), you can get that listing with the `what` command.

What is the MATLAB path? MATLAB path is a variable, stored under the name `path` that contains the paths of all directories that are automatically included in MATLAB's search path. By default, all directories that are installed by the MATLAB installer are included in this path. MATLAB 6 includes a default **work** directory where MATLAB takes you automatically when you launch MATLAB. So, if you don't do any directory navigation, all files that you create and save during a MATLAB session will be saved in the **work** directory. However, you are not limited to this directory for saving your files. You can create a directory

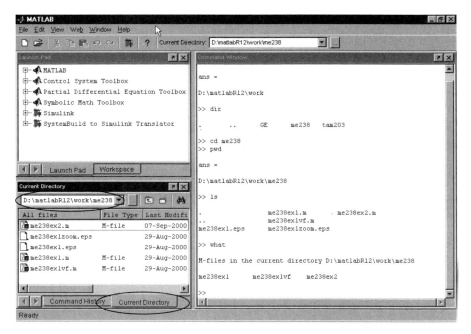

Figure 2.8: The MATLAB desktop.

anywhere you like on the computer and access it as mentioned above in *How do you change the current directory.* This change, however, is effective only for the duration of the current session.

If you prefer to organize your MATLAB files in different directories (other than **work**) and would like to have access to all your files automatically each time you work in MATLAB, you should modify the MATLAB search path with the `path` or `addpath` command to include your directories in the search path.

Example:

- Create two new directories: (i) a directory called **tutorials** inside the **work** directory, and (ii) a directory called **mywork** on **C:** drive.

- Use command `addpath` to add these directories to the existing path.

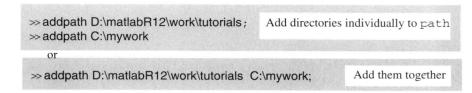

- Type `path` on the command line to see the new MATLAB search path and check if your directories are included (right on the top of the stack).

- You can also use `path` command to add these directories to the search path. For example, `path=path(path,'C:\mywork')` adds the directory `C:\mywork` to the current search path. This command, however, is likely to be confusing to a beginner in MATLAB.

- Alternatively, you can navigate and get your desired directory to the **Current Directory** subwindow, right-click the mouse inside that subwindow to bring up a pull down menu, and select **add to path** to add the directory to the search path.

What are those other windows? While you are at it, explore the other subwindows of the MATLAB desktop. On the left of the **Command Window**, there are four other subwindows, usually two visible and two hidden behind (see Section 1.6.1 on page 8 and Fig. 1.3 on page 10). In particular, do the following.

- **Command History** subwindow: Click on the tab for this subwindow if it is not already in the foreground. Type the following commands in the command window.

```
t = linspace(0,2*pi,100);   % take 100 points in [0,2*pi]
x = cos(t);
y = sin(t);
plot(x,y)                   % this is a unit circle
```

  - Now, go to the **Command History** window and double click on any of the commands listed to execute it again in the command window.

  - Select all the four commands (shown above) in the **Command History** window, right click the mouse, and select **Create M-File**. Thus you can test commands and turn them into M-files easily.

- **Workspace** subwindow: Click on the tab for this subwindow if it is not already in the foreground.

  - Select any of the variables, `t`, `x`, or `y`, listed in this subwindow and double click to open an **Array Editor** window. You can change any values in the array here.

  - Select any variable and right click the mouse. Select any option from the pop-up menu. You can edit, import, graph, or save the variable with a click of the mouse.

3. *Interactive Computation*

In principle, one can do all calculations in MATLAB interactively, by entering commands sequentially in the command window, although a *script file* (explained in Section 4.1) is perhaps a better choice for computations that involve more than a few steps. The interactive mode of computation, however, makes MATLAB a powerful scientific calculator that puts hundreds of built-in mathematical functions for numerical calculations and sophisticated graphics at the finger tips of the user.

In this chapter, we introduce you to some of MATLAB's built-in functions and capabilities, through examples of interactive computation. The basic things to keep in mind are:

Where to type commands: All MATLAB commands or expressions are entered in the command window at the MATLAB prompt '≫ '.

How to execute commands: To execute a command or statement, you must press return or enter at the end.

What to do if the command is very long: If your command does not fit on one line you can continue the command on the next line if you type three consecutive periods at the end of the first line. You can keep continuing this way till the length of your command hits the limit, which is 4096 characters. For more information see the discussion on **Continuation** on pages 49 and 97.

How to name variables: Names of variables must begin with a letter. After the first letter, any number of digits or underscores may be used, but MATLAB remembers only the first 31 characters.

What is the precision of computation: All computations are carried out internally in double precision. The appearance of numbers on the screen, however, depends on the `format` in use (see Section 1.6.3).

How to control the display format of output: The output appearance of floating point numbers (number of digits after the decimal, etc.) is controlled with the `format` command. The default is `format short`, which displays four digits after the decimal. For other available formats and how to change them, see Section 1.6.3 on page 9 or on-line help on `format` .

How to suppress the screen output: A semicolon (;) at the end of a command suppresses the screen output, although the command is carried out and the result is saved in the variable assigned to the command or in the default variable `ans`.

How to set paged-screen display: For paged-screen display (one screenful of output display at a time) use the command `more on`.

Where and how to save results: If you need to save some of the computed results for later processing, you can save the variables in a file in binary or ASCII format with the `save` command. See Section 3.5 on page 74 for more information.

How to print your work: You can print your entire session in MATLAB, part of it, or selected segments of it, in one of several ways. The simplest way, perhaps, is to create a diary with the `diary` command (see Section 3.5.3 for more information) and save your entire session in it. Then you can print the diary just the way you would print any other file on your computer. On PCs and Macs, however, you can print the session by selecting **Print** from the **File** menu. (Before you print, make sure that the command window is the active window. If it isn't, just click on the command window to make it active).

What about comments: MATLAB takes anything following a `%` as a comment and ignores it [1]. You are not likely to use a lot of comments while computing interactively, but you will use them when you write programs in MATLAB.

Since MATLAB derives most of its power from matrix computations and assumes every variable to be, at least potentially, a matrix, we start with descriptions and examples of how to enter, index, manipulate, and perform some useful calculations with matrices.

---

[1] except when the `%` appears in a quote enclosed character string or in certain I/O format statements.

3.1   Matrices and Vectors

3.1.1   Input

For on-line help, type: `help elmat`

A matrix is entered row-wise, with consecutive elements of a row separated by a space or a comma, and the rows separated by semicolons or carriage returns. The entire matrix must be enclosed within square brackets. Elements of the matrix may be real numbers, complex numbers, or valid MATLAB expressions.

Examples:

Matrix	**MATLAB input command**
$A = \begin{bmatrix} 1 & 2 & 5 \\ 3 & 9 & 0 \end{bmatrix}$	`A = [1 2 5; 3 9 0]`
$B = \begin{bmatrix} 2x & \ln x + \sin y \\ 5i & 3 + 2i \end{bmatrix}$	`B = [2*x log(x)+sin(y); 5i 3+2i]`

Special cases: vectors and scalars

- A vector is a special case of a matrix, with just one row or one column. It is entered the same way as a matrix.
  Examples: `u = [1 3 9]` produces a row vector,
  `v = [1; 3; 9]` produces a column vector.

- A scalar does not need brackets.
  Example: `g = 9.81;`

- Square brackets with no elements between them create a null matrix.
  Example: `X = [].` (See Fig. 3.1 for a more useful example).

Continuation

If it is not possible to type the entire input on the same line then use three consecutive periods (...) to signal continuation, and continue the input on the next line. The three periods are called an *ellipsis.* For example,

```
A = [1/3    5.55*sin(x)   9.35   0.097;...
     3/(x+2*log(x))   3    0    6.555; ...
     (5*x-23)/55    x-3    x*sin(x)    sqrt(3)];
```

produces the intended 3×4 matrix A (provided, of course, x has been assigned a value before). A matrix can also be entered across multiple lines using carriage returns at the end of each row. In this case, the semicolons and ellipses at the end of each row may be omitted. Thus, the following three commands are equivalent:

---

[2]This box lets you know that you can learn more about this topic from MATLAB's online help, in the help category called *elmat* (for *el*ementary *mat*rix manipulations).

```
A = [1 3 9; 5 10 15; 0 0 -5];
A = [1 3 9
     5 10 15
     0 0 -5];
A = [1 3 9; 5 10 ...
     15; 0 0 -5];
```

Continuation across several input lines achieved through '...' is not limited to matrix input. This construct may be used for other commands and for a long list of command arguments (see Section 4.3.2 on page 97), as long as the command does not exceed 4096 characters.

3.1.2 Indexing (or Subscripting)

Once a matrix exists, its elements are accessed by specifying their row and column indices. Thus `A(i,j)` in MATLAB refers to the element a_{ij} of matrix A, i.e., the element in the ith row and jth column. This notation is fairly common in computational software packages and programming languages. MATLAB, however, provides a much higher level of index specification — it allows a range of rows and columns to be specified at the same time. For example, the statement `A(m:n,k:l)` specifies rows m to n and columns k to l of matrix A. When the rows (or columns) to be specified range over all rows (or columns) of the matrix, a colon can be used as the row (or column) index. Thus `A(:,5:20)` refers to the elements in columns 5 through 20 of *all* the rows of matrix A. This feature makes matrix manipulation much easier and provides a way to take advantage of the 'vectorized' nature of calculations in MATLAB. (See Fig. 3.1 on page 51 for examples).

Dimension

Matrix dimensions are determined automatically by MATLAB, i.e., no explicit dimension declarations are required. The dimensions of an existing matrix A may be obtained with the command `size(A)` or more explicitly with `[m,n] = size(A)`, which assigns the number of rows and columns of A to the variables m and n, respectively. When a matrix is entered by specifying a single element or a few elements of the matrix, MATLAB creates a matrix just big enough to accommodate the elements. Thus if the matrices B and C do not exist already, then

$$B(2,3) = 5; \qquad \text{produces} \qquad B = \begin{bmatrix} 0 & 0 & 0 \\ 0 & 0 & 5 \end{bmatrix},$$

$$C(3,1:3) = [1\ 2\ 3]; \quad \text{produces} \qquad C = \begin{bmatrix} 0 & 0 & 0 \\ 0 & 0 & 0 \\ 1 & 2 & 3 \end{bmatrix}.$$

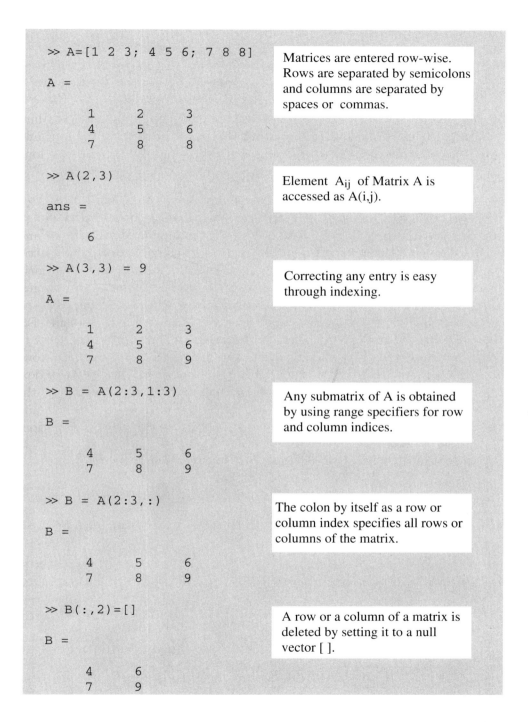

```
>> A=[1 2 3; 4 5 6; 7 8 8]

A =

        1        2        3
        4        5        6
        7        8        8
```
Matrices are entered row-wise. Rows are separated by semicolons and columns are separated by spaces or commas.

```
>> A(2,3)

ans =

        6
```
Element A_{ij} of Matrix A is accessed as A(i,j).

```
>> A(3,3) = 9

A =

        1        2        3
        4        5        6
        7        8        9
```
Correcting any entry is easy through indexing.

```
>> B = A(2:3,1:3)

B =

        4        5        6
        7        8        9
```
Any submatrix of A is obtained by using range specifiers for row and column indices.

```
>> B = A(2:3,:)

B =

        4        5        6
        7        8        9
```
The colon by itself as a row or column index specifies all rows or columns of the matrix.

```
>> B(:,2)=[]

B =

        4        6
        7        9
```
A row or a column of a matrix is deleted by setting it to a null vector [ ].

Figure 3.1: Examples of matrix input and matrix index manipulation.

3.1.3 Matrix Manipulation

As you can see from examples in Fig. 3.1 on page 51, it is fairly easy to correct wrong entries of a matrix, extract any part or submatrix of a matrix, or delete or add rows and columns. These manipulations are done with MATLAB's smart indexing feature. By specifying vectors as the row and column indices of a matrix one can reference and modify any submatrix of a matrix. Thus if A is a 10×10 matrix, B is a 5×10 matrix, and y is a 20 elements long row vector, then

$$A([1\ 3\ 6\ 9],:) \quad = \quad [B(1:3,:);\ y(1:10)]$$

replaces 1st, 3rd, and 6th rows of A by the first 3 rows of B, and the 9th row of A by the first 10 elements of y. In such manipulations, it is imperative, of course, that the sizes of the submatrices to be manipulated are compatible. For example, in the above assignment, number of columns in A and B must be the same, and the total number of rows specified on the right hand side must be the same as the number of rows specified on the left.

In MATLAB 4.x, you could use a sophisticated way of indexing with 0-1 vectors (usually created by relational operations) to reference submatrices. For example, let v = [1 0 0 1 1] and Q be a 5×5 matrix. Then Q(v,:) picks out those rows of Q where v is non-zero, i.e., the 1st, 4th, and 5th rows. In MATLAB 5.x, this can be done directly by creating index vectors with numbers representing the desired rows or columns.

$$\text{So, if} \quad Q = \begin{bmatrix} 2 & 3 & 6 & 0 & 5 \\ 0 & 0 & 20 & -4 & 3 \\ 1 & 2 & 3 & 9 & 8 \\ 2 & -5 & 5 & -5 & 6 \\ 5 & 10 & 15 & 20 & 25 \end{bmatrix} \quad \text{and} \quad v = \begin{bmatrix} 1 & 4 & 5 \end{bmatrix},$$

$$\text{then,} \quad Q(v,:) = \begin{bmatrix} 2 & 3 & 6 & 0 & 5 \\ 2 & -5 & 5 & -5 & 6 \\ 5 & 10 & 15 & 20 & 25 \end{bmatrix}, \text{ and } Q(:,v) = \begin{bmatrix} 2 & 0 & 5 \\ 0 & -4 & 3 \\ 1 & 9 & 8 \\ 2 & -5 & 6 \\ 5 & 20 & 25 \end{bmatrix}.$$

In MATLAB 5 onwards, matrix indexing with 0-1 vectors can be done if

1. The vector is produced by logical or relational operations (see Sections 3.2.2 and 3.2.3 on pages 59 and 59).

2. The 0-1 vector created by you is converted into a logical array with the command `logical`. For example, to get the 1st, 4th, and 5th rows of Q with 0-1 vectors, you can do:

```
v = [1 0 0 1 1];     v = logical(v);     Q(v,:).
```

Reshaping matrices

Matrices can be reshaped into a vector or any other appropriately sized matrix:

As a vector: All the elements of a matrix A can be strung into a single column vector b by the command b = A(:) (matrix A is stacked in vector b columnwise).

As a differently sized matrix: If matrix A is an $m \times n$ matrix, it can be reshaped into a $p \times q$ matrix, as long as $m \times n = p \times q$, with the command reshape(A,p,q). Thus, for a 6×6 matrix A,

reshape(A,9,4)   transforms A into a 9×4 matrix,
reshape(A,3,12)   transforms A into a 3×12 matrix.

Now let us look at some frequently used manipulations.

Transpose

The transpose of a matrix A is obtained by typing A', i.e., the name of the matrix followed by the single right quote. For a real matrix A, the command B = A' produces $B = A^T$, that is, $b_{ij} = a_{ji}$, and for a complex matrix A, B = A' produces the conjugate transpose $B = \bar{A}^T$, that is, $b_{ij} = \bar{a}_{ji}$.

Examples:

If $A = \begin{bmatrix} 2 & 3 \\ 6 & 7 \end{bmatrix}$,   then B = A' gives $B = \begin{bmatrix} 2 & 6 \\ 3 & 7 \end{bmatrix}$.

If $C = \begin{bmatrix} 2 & 3+i \\ 6i & 7i \end{bmatrix}$,   then Ct = C' gives $Ct = \begin{bmatrix} 2 & -6i \\ 3-i & -7i \end{bmatrix}$.

If $u = [0 \ 1 \ 2 \ \cdots \ 9]$,   then v = u(3:6)' gives $v = \begin{bmatrix} 2 \\ 3 \\ 4 \\ 5 \end{bmatrix}$.

Initialization

Initialization of a matrix is not necessary in MATLAB. However, it is advisable in the following two cases.

1. **Large matrices:** If you are going to generate or manipulate a large matrix, initialize the matrix to a zero matrix of the required dimension. An $m \times n$ matrix can be initialized by the command A = zeros(m,n). The initialization reserves for the matrix a contiguous block in the computer's memory. Matrix operations performed on such matrices are generally more efficient.

2. **Dynamic matrices:** If the rows or columns of a matrix are computed in a loop (e.g. `for` or `while` loop) and appended to the matrix (see below) in each execution of the loop, then you might want to initialize the matrix to a null matrix before the loop starts. A null matrix A is created by the command `A = []`. Once created, a row or column of any size may be appended to A as described below.

Appending a row or column

A row can be easily appended to an existing matrix provided the row has the same length as the length of the rows of the existing matrix. The same thing goes for columns. The command `A = [A u]` appends the column vector u to the columns of A, while `A = [A; v]` appends the row vector v to the rows of A. A row or column of any size may be appended to a null matrix.

Examples: If

$$A = \begin{bmatrix} 1 & 0 & 0 \\ 0 & 1 & 0 \\ 0 & 0 & 1 \end{bmatrix}, \quad u = \begin{bmatrix} 5 & 6 & 7 \end{bmatrix}, \text{ and } v = \begin{bmatrix} 2 \\ 3 \\ 4 \end{bmatrix},$$

then

A = [A; u] produces $\quad A = \begin{bmatrix} 1 & 0 & 0 \\ 0 & 1 & 0 \\ 0 & 0 & 1 \\ 5 & 6 & 7 \end{bmatrix}, \quad$ a 4×3 matrix,

A = [A v] produces $\quad A = \begin{bmatrix} 1 & 0 & 0 & 2 \\ 0 & 1 & 0 & 3 \\ 0 & 0 & 1 & 4 \end{bmatrix}, \quad$ a 3×4 matrix,

A = [A u'] produces $\quad A = \begin{bmatrix} 1 & 0 & 0 & 5 \\ 0 & 1 & 0 & 6 \\ 0 & 0 & 1 & 7 \end{bmatrix}, \quad$ a 3×4 matrix,

A = [A u] produces $\quad$ an error,

B = []; B = [B; 1 2 3] produces $\quad B = \begin{bmatrix} 1 & 2 & 3 \end{bmatrix}$, and

B=[]; for k=1:3, B=[B; k k+1 k+2]; end produces $\quad B = \begin{bmatrix} 1 & 2 & 3 \\ 2 & 3 & 4 \\ 3 & 4 & 5 \end{bmatrix}.$

Deleting a row or column

Any row(s) or column(s) of a matrix can be deleted by setting the row or column to a null vector.

Examples:

A(2,:)   = []   deletes the 2nd row of matrix A,
A(:,3:5) = []   deletes the 3rd through 5th columns of A,

```
A([1 3],:) = []          deletes the 1st and the 3rd row of A,
u(5:length(u)) = []      deletes all elements of vector u except 1 through 4.
```

For on-line help type: `help elmat`

Utility matrices

To aid matrix generation and manipulation, MATLAB provides many useful utility matrices. For example,

`eye(m,n)`	returns an m by n matrix with 1's on the main diagonal
`zeros(m,n)`	returns an m by n matrix of zeros
`ones(m,n)`	returns an m by n matrix of ones
`rand(m,n)`	returns an m by n matrix of random numbers
`randn(m,n)`	returns an m by n matrix of normally distributed numbers
`diag(v)`	generates a diagonal matrix with vector v on the diagonal
`diag(A)`	extracts the diagonal of matrix A as a vector
`diag(A,1)`	extracts the first upper off-diagonal vector of matrix A.

The first four commands with a single argument, e.g. `ones(m)`, produce square matrices of dimension m. For example, `eye(3)` produces a 3×3 identity matrix. A matrix can be built with many block matrices as well. See examples in Fig. 3.2.

Here is a list of some more functions used in matrix manipulation:

`rot90`	rotates a matrix by 90°
`fliplr`	flips a matrix from left to right
`flipud`	flips a matrix from up to down
`tril`	extracts the lower triangular part of a matrix
`triu`	extracts the upper triangular part of a matrix
`reshape`	changes the shape of a matrix.

Special matrices

For on-line help type: `help specmat`

There is also a set of built-in special matrices such as `hadamard`, `hankel`, `hilb`, `invhilb`, `kron`, `pascal`, `toeplitz`, `vander`, `magic`, etc. For a complete list and help on these matrices, type `help specmat`.

3.1.4   Creating Vectors

Very often we need to create a vector of numbers over a given range with a specified increment. The general command to do this in MATLAB is

$$\boxed{\texttt{v = }\textit{InitialValue}\texttt{ : }\textit{Increment}\texttt{ : }\textit{FinalValue}}$$

The three values in the above assignment can also be valid MATLAB expressions. If no increment is specified, MATLAB uses the default increment of 1.

Examples:

```
>> eye(3)

ans =

     1     0     0
     0     1     0
     0     0     1
```

eye(n) creates an n by n identity matrix. The commands zeros, ones, and rand work in a similar way.

```
>> B = [ones(3) zeros(3,2); zeros(2,3) 4*eye(2)]

B =

     1     1     1     0     0
     1     1     1     0     0
     1     1     1     0     0
     0     0     0     4     0
     0     0     0     0     4
```

Create a matrix B using submatrices made up of elementary matrices: ones, zeros, and the identity matrix of the specified sizes.

```
>> diag(B)'

ans =

     1     1     1     4     4
```

This command pulls out the diagonal of B in a row vector. Without the transpose, the result would obviously be a column vector.

```
>> diag(B,1)'

ans =

     1     1     0     0
```

The second argument of the command specifies the off-diagonal vector to be pulled out. Here we get the first upper off-diagonal vector. A negative value of the argument specifies the lower off-diagonal vectors.

```
>> d = [2 4 6 8];
>> d1 = [-3 -3 -3];
>> d2 = [-1 -1];
>> D = diag(d) + diag(d1,1) + diag(d2,-2)
```

Create vectors d, d1, and d2 of length 4, 3, and 2 respectively.

```
D =

     2    -3     0     0
     0     4    -3     0
    -1     0     6    -3
     0    -1     0     8
```

Create a matrix D by putting d on the main diagonal, d1 on the first upper diagonal and d2 on the second lower diagonal.

Figure 3.2: Examples of matrix manipulation using utility matrices and functions.

`a = 0:10:100`	produces $a = [\ 0 \quad 10 \quad 20 \quad \ldots \quad 100\ ]$,
`b = 0:pi/50:2*pi`	produces $b = [\ 0 \quad \frac{\pi}{50} \quad \frac{2\pi}{50} \quad \ldots \quad 2\pi\ ]$, i.e., a linearly spaced vector from 0 to 2π spaced at $\pi/50$,
`u = 2:10`	produces $a = [\ 2 \quad 3 \quad 4 \quad \ldots \quad 10\ ]$.

As you may notice, no square brackets are required if a vector is generated this way, however, a vector assignment such as `u = [1:10 33:-2:19]` does require square brackets to force the concatenation of the two vectors: `[1 2 3 ... 10]` and `[33 31 29 ... 19]`. Finally, we mention the use of two frequently used built-in functions to generate vectors:

`linspace(a,b,n)` generates a linearly spaced vector of length n from a to b.
    Example: `u=linspace(0,20,5)` generates `u=[0 5 10 15 20]`.
    Thus `u=linspace(a,b,n)` is the same as `u=a:(b-a)/(n-1):b`.

`logspace(a,b,n)` generates a logarithmically spaced vector of length n from 10^a to 10^b.
    Example: `v=logspace(0,3,4)` generates `v=[1 10 100 1000]`.
    Thus `logspace(a,b,n)` is the same as `10.^(linspace(a,b,n))`. (The array operation `.^` is discussed in the next section.)

Special vectors, such as vectors of zeros or ones of a specific length, can be created with the utility matrix functions `zeros`, `ones`, etc.

Examples:

`u = zeros(1,1000)`	initializes a 1000 element long row vector
`v = ones(10,1)`	creates a 10 element long column vector of 1's.

3.2   Matrix and Array Operations

For on-line help type:
`help ops`

3.2.1   Arithmetic operations

For people who are used to programming in a conventional language like Pascal, Fortran, or C, it is an absolute delight to be able to write a matrix product as `C = A*B` where `A` is an $m \times n$ matrix and B is an $n \times k$ matrix [2]. MATLAB allows all arithmetic operations:

For on-line help type:
`help arith`

$+$		addition
$-$		subtraction
$*$		multiplication
$/$		division
$\hat{\ }$	(caret)	exponentiation

to be carried out on matrices in straightforward ways as long as the operation makes sense mathematically and the operands are compatible. Thus,

---

[2]although you can do `C = A*B` in C^{++}.

A+B or A−B   is valid if A and B are of the same size,
A∗B   is valid if A's number of columns equals B's number of rows,
A/B   is valid, and equals $A \cdot B^{-1}$ for same-size square matrices A & B,
A^2   makes sense only if A is square, and equals A∗A.

In all the above commands if B is replaced by a scalar, say α, the arith-metic operations are still carried out. In this case, the command A+α adds α to each element of A, the command A∗α (or α∗A) multiplies each element of A by α and so on. Vectors, of course, are just treated as a single row or a column matrix and therefore a command such as w = u∗v, where u and v are same size vectors, say $m \times 1$, produces an error (because you cannot multiply an $m \times 1$ matrix with an $m \times 1$ matrix) while w = u∗v′ and w = u′∗v execute correctly, producing the outer and the inner products of the two vectors, respectively (see examples in Fig 3.3).

The left division: In addition to the normal or *right* division ($/$), there is a *left* division ($\backslash$) in MATLAB. This division is used to solve a matrix equation. In particular, the command x = A\b solves the matrix equation **A** x = **b**. Thus A\b is *almost* the same as inv(A)∗b, but faster and more numerically stable than computing inv(A)∗b. In the degenerate case of scalars 5\3 gives 0.6, which is $3/5$ or $5^{-1} * 3$.

Array operation:

How does one get products like $[u_1v_1 \ \ u_2v_2 \ u_3v_3 \ \ldots \ u_nv_n]$ from two vectors u and v? No, you do not have to use *DO* or *FOR* loops. You can do *array operation* — operations done on element-by-element basis.   Element-by-element multiplication, division, and exponentiation between two matrices or vectors of the same size are done by preceding the corresponding arithmetic operators by a period (.):

.∗	element-by-element multiplication
./	element-by-element left division
.\	element-by-element right division
.^	element-by-element exponentiation
.′	nonconjugated transpose

Examples:

u.∗v   produces $[u_1v_1 \ \ u_2v_2 \ u_3v_3 \ \ldots],$
u./v   produces $[u_1/v_1 \ \ u_2/v_2 \ u_3/v_3 \ \ldots],$ and
u.^v   produces $[u_1^{v_1}, \ u_2^{v_2}, \ u_3^{v_3}, \ \ldots].$

The same is true for matrices. For two same-sized matrices A and B, the command C = A.∗B produces a matrix C with elements $C_{ij} = A_{ij}B_{ij}$. Clearly, there is a big difference between A^2 and A.^2 (see Fig. 3.3). Once again, scalars do enjoy a special status. While `u./v` or `u.^v` will produce an error if u and v are not of the same size, `1./v` happily computes $[1/v_1 \ 1/v_2 \ 1/v_3 \ \ldots]$, and `pi.^v` gives $[\pi^{v_1} \ \pi^{v_2} \ \pi^{v_3} \ \ldots]$.

3.2.2 Relational operations

For on-line help type:
`help relop`

There are six relational operators in MATLAB:

$<$	less than
$<=$	less than or equal
$>$	greater than
$>=$	greater than or equal
$==$	equal
$\tilde{\ }=$	not equal.

These operations result in a vector or matrix of the same size as the operands, with 1 where the relation is true and 0 where it is false.

Examples: If $x = [1 \ 5 \ 3 \ 7]$ and $y = [0 \ 2 \ 8 \ 7]$, then

`k = x < y`	results in $k = [0 \ 0 \ 1 \ 0]$	because $x_i < y_i$ for $i = 3$,
`k = x <= y`	results in $k = [0 \ 0 \ 1 \ 1]$	because $x_i \leq y_i$ for $i = 3$ and 4,
`k = x > y`	results in $k = [1 \ 1 \ 0 \ 0]$	because $x_i > y_i$ for $i = 1$ and 2,
`k = x >= y`	results in $k = [1 \ 1 \ 0 \ 1]$	because $x_i \geq y_i$ for $i = 1, 2$ and 4,
`k = x == y`	results in $k = [0 \ 0 \ 0 \ 1]$	because $x_i = y_i$ for $i = 4$, and
`k = x ~= y`	results in $k = [1 \ 1 \ 1 \ 0]$	because $x_i \neq y_i$ for $i = 1, 2$ and 3.

Although these operations are usually used in conditional statements such as *if-then-else* to branch out to different cases, they can be used to do pretty sophisticated matrix manipulation. For example, `u = v(v >= sin(pi/3))` finds all elements of vector v such that $v_i \geq \sin\frac{\pi}{3}$ and stores them in vector u. Two or more of these operations can also be combined with the help of *logical operators* (described below).

3.2.3 Logical operations

For on-line help type:
`help relop`

There are four logical operators:

`&`	logical AND
`|`	logical OR
`~`	logical complement (NOT)
`xor`	exclusive OR

These operators work in a similar way as the relational operators and produce vectors or matrices of the same size as the operands, with 1 where the condition is true and 0 where false.

Examples: For two vectors $x = [0 \ 5 \ 3 \ 7]$ and $y = [0 \ 2 \ 8 \ 7]$,

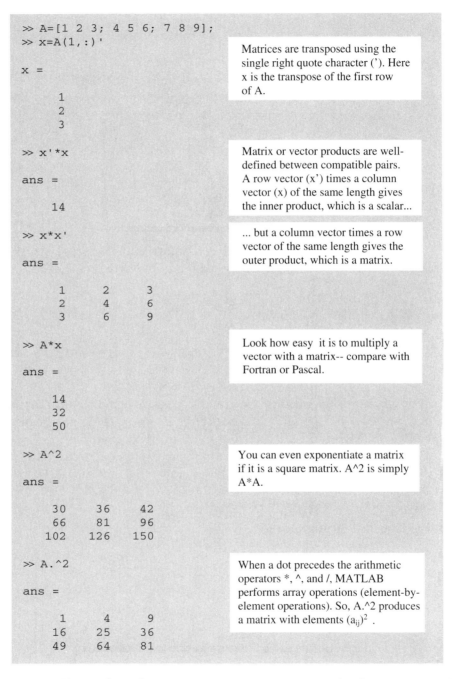

```
>> A=[1 2 3; 4 5 6; 7 8 9];
>> x=A(1,:)'

x =

     1
     2
     3
```

Matrices are transposed using the single right quote character ('). Here x is the transpose of the first row of A.

```
>> x'*x

ans =

    14
```

Matrix or vector products are well-defined between compatible pairs. A row vector (x') times a column vector (x) of the same length gives the inner product, which is a scalar...

```
>> x*x'

ans =

     1     2     3
     2     4     6
     3     6     9
```

... but a column vector times a row vector of the same length gives the outer product, which is a matrix.

```
>> A*x

ans =

    14
    32
    50
```

Look how easy it is to multiply a vector with a matrix-- compare with Fortran or Pascal.

```
>> A^2

ans =

    30    36    42
    66    81    96
   102   126   150
```

You can even exponentiate a matrix if it is a square matrix. A^2 is simply A*A.

```
>> A.^2

ans =

     1     4     9
    16    25    36
    49    64    81
```

When a dot precedes the arithmetic operators *, ^, and /, MATLAB performs array operations (element-by-element operations). So, A.^2 produces a matrix with elements $(a_{ij})^2$.

Figure 3.3: Examples of matrix transpose, matrix multiplication, matrix exponentiation, and array exponentiation.

m = (x>y)&(x>4)	results in m = [0 1 0 0], since the condition is true only for x_2.
n = x|y	results in n = [0 1 1 1], since either x_i or y_i is non-zero for $i = 2, 3$ and 4.
m = ~(x|y)	results in m = [1 0 0 0], which is the logical complement of x|y.
p = xor(x,y)	results in p = [0 0 0 0], since there is no such index i for which x_i or y_i, but not both, is non-zero.

Since the output of the logical operations is a 0-1 vector or a 0-1 matrix, the output can be used as the index of a matrix to extract appropriate elements. For example, to see those elements of x that satisfy both the conditions (x > y) & (x > 4), type x((x>y)&(x>4)).

In addition to these logical operators, there are many useful built-in logical functions, such as: :

all	true (= 1) if all elements of a vector are true. *Example:* all(x<0) returns 1 if each element of x is negative.
any	true (= 1) if any element of a vector is true. *Example:* any(x) returns 1 if any element of x is non-zero.
exist	true (= 1) if the argument (a variable or a function) exists.
isempty	true (= 1) for an empty matrix.
isinf	true for all infinite elements of a matrix.
isfinite	true for all finite elements of a matrix.
isnan	true for all elements of a matrix that are Not-A-Number.[3]
find	finds indices of non-zero elements of a matrix. *Examples:* find(x) returns [2 3 4] for x=[0 2 5 7], i=find(x>5) returns $i = 4$. [r,c] = find(A>100) returns the row and column indices i and j of A, in vectors r and c, for which $A_{ij} > 100$.

To complete this list of logical functions, we just mention isreal, issparse, isstr, and isglobal.

3.2.4 Elementary math functions

For on-line help type:
help elfun

All of the following built-in math functions take matrix inputs and perform array operations (element-by-element) on them. Thus, they produce an output matrix of the same size as the input matrix.

Trigonometric functions

sin	Sine.	sinh	Hyperbolic sine.
asin	Inverse sine.	asinh	Inverse hyperbolic sine.
cos	Cosine.	cosh	Hyperbolic cosine.
acos	Inverse cosine.	acosh	Inverse hyperbolic cosine.
tan	Tangent.	tanh	Hyperbolic tangent.
atan,atan2	Inverse tangent.	atanh	Inverse hyperbolic tangent.
sec	Secant.	sech	Hyperbolic secant.
asec	Inverse secant.	asech	Inverse hyperbolic secant.
csc	Cosecant.	csch	Hyperbolic cosecant.

acsc	Inverse cosecant.	acsch	Inverse hyperbolic cosecant.
cot	Cotangent.	coth	Hyperbolic cotangent.
acot	Inverse cotangent.	acoth	Inverse hyperbolic cotangent

The angles given to these functions as arguments must be in radians. All of these functions, except `atan2`, take a single scalar, vector, or matrix as input argument. The function `atan2` takes two input arguments: `atan2(y,x)` and produces the four quadrant inverse tangent such that $-\pi \leq \tan^{-1}\frac{y}{x} \leq \pi$. This gives the angle in a rectangular to polar conversion.

Examples: If `q=[0 pi/2 pi]`, `x=[1 -1 -1 1]`, and `y=[1 1 -1 -1]`, then

> `sin(q)`   gives `[0 1 0]`,
> `sinh(q)`   gives `[0 2.3013 11.5487]`,
> `atan(y./x)` gives `[0.7854 -0.7854 0.7854 -0.7854]`, and
> `atan2(y,x)` gives `[0.7854 2.3562 -2.3562 -0.7854]`.

Exponential functions

exp	Exponential.
	Example: `exp(A)` produces a matrix with elements $e^{(A_{ij})}$.
	So how do you compute e^A? See the next section.
log	Natural logarithm.
	Example: `log(A)` produces a matrix with elements $\ln(A_{ij})$.
log10	Base 10 logarithm.
	Example: `log10(A)` produces a matrix with elements $\log_{10}(A_{ij})$.
sqrt	Square root.
	Example: `sqrt(A)` produces a matrix with elements $\sqrt{A_{ij}}$.
	But what about $\sqrt{A}$? See the next section.

In addition, `log2` , `pow2`, and `nextpow2` functions have been added in MATLAB 6. Clearly, these are array operations. You can, however, also compute matrix exponential e^A, matrix square root $\sqrt{A}$, etc. See the next section (Section 3.2.5).

Complex functions

abs	Absolute value.
	Example: `abs(A)` produces a matrix of absolute values $|A_{ij}|$.
angle	Phase angle.
	Example: `angle(A)` gives the phase angles of complex A.
conj	Complex conjugate.
	Example: `conj(A)` produces a matrix with elements $\bar{A}_{ij}$.
imag	Imaginary part.
	Example: `imag(A)` extracts the imaginary part of A.
real	Real part.
	Example: `real(A)` extracts the real part of A.

Round-off functions

fix	Round towards 0.
	Example: fix([-2.33 2.66]) = $[-2 \quad 2]$.
floor	Round towards $-\infty$.
	Example: floor([-2.33 2.66]) = $[-3 \quad 2]$.
ceil	Round towards $+\infty$.
	Example: ceil([-2.33 2.66]) = $[-2 \quad 3]$.
round	Round towards the nearest integer.
	Example: round([-2.33 2.66]) = $[-2 \quad 3]$.
rem	Remainder after division. rem(a,b) is the same as a - fix(a./b).
	Example: If a=[-1.5 7], b=[2 3], then rem(a,b) = [-1.5 1].
sign	Signum function.
	Example: sign([-2.33 2.66]) = $[-1 \quad 1]$.

3.2.5   Matrix functions

For on-line help type:
help matfun

We discussed the difference between the array exponentiation A.^2 and the matrix exponentiation A^2 above. There are some built-in functions that are truly matrix functions, and that also have array counterparts. The matrix functions are:

expm(A)	finds the exponential of matrix A, e^A,
logm(A)	finds $\log(A)$ such that $A = e^{\log(A)}$,
sqrtm(A)	finds $\sqrt{A}$.

The array counterparts of these functions are **exp, log,** and **sqrt,** which operate on each element of the input matrix (see Fig. 3.4 for examples). The matrix exponential function **expm** also has some specialized variants **expm1, expm2,** and **expm3.** See the on-line help or the Reference Guide [2] for their proper usage. MATLAB also provides a general matrix function **funm** for evaluating true matrix functions.

3.2.6   Character strings

For on-line help type:
help strfun.

All character strings are entered within two single right quote characters— *'string'.* MATLAB treats every string as a row vector with one element per character. For example, typing

message = 'Leave me alone'

creates a vector, named **message,** of size 1×14 (spaces in strings count as characters). Therefore, to create a column vector of text objects, each text string must have exactly the same number of characters. For example, the command

names = ['John'; 'Ravi'; 'Mary'; 'Xiao']

creates a column vector with one name per row, although, to MATLAB, the variable **names** is a 4×4 matrix. Clearly, the command howdy = ['Hi'; 'Hello'; 'Namaste'] will result in an error because each row has different

```
>> A=[1 2; 3 4];

A =

     1      2
     3      4

>> asqrt = sqrt(A)

asqrt =

    1.0000     1.4142
    1.7321     2.0000
```

sqrt is an array operation. It gives the square root of each element of A as is evident from the output here.

```
>> Asqrt = sqrtm(A)

Asqrt =

   0.5537 + 0.4644i    0.8070 - 0.2124i
   1.2104 - 0.3186i    1.7641 + 0.1458i
```

sqrtm, on the other hand, is a true matrix function, i.e., it computes $\sqrt{A}$. Thus [Asqrt]*[Asqrt] = [A].

```
>> exp_aij = exp(A)

exp_aij =

    2.7183     7.3891
   20.0855    54.5982
```

Similarly, exp gives element-by-element exponential of the matrix, whereas expm finds the true matrix exponential e^A. For info on other matrix functions, type help matfun.

```
>> exp_A = expm(A)

exp_A =

   51.9690    74.7366
  112.1048   164.0738
```

Figure 3.4: Examples of differences between matrix functions and array functions.

length. Text strings of different lengths can be made to be of equal length by padding them with blanks. Thus the correct input for `howdy` will be
`howdy = ['Hi␣␣␣␣␣';'Hello␣␣'; 'Namaste']`
with each string being 7 characters long (␣ denotes a space).

An easier way of doing the same thing is to use the command `char`, which converts strings to a matrix. `char(s1,s2,...)` puts each string argument `s1`, `s2`, etc., in a row and creates a string matrix by padding each row with the appropriate number of blanks. Thus, to create the same `howdy` as above, we type

`howdy = char('Hi','Hello','Namaste').`

The function `char` is a MATLAB 5.x feature. In MATLAB 4.2x, the equivalent function to create a string array is `str2mat`. This function still exists in MATLAB 5.x but would probably become obsolete in future versions.

Since the same quote character (the single right quote) is used to signal the beginning as well as the end of a string, this character cannot be used inside a string for a quote or apostrophe. For a quote within a string you must use the double quote character ("). Thus, to title a graph with *3-D View of Boomerang's Path* you write

`title('3-D View of Boomerang"s Path').`

Manipulating character strings

Character strings can be manipulated just like matrices. Thus,

`c = [howdy(2,:)  names(3,:)]`

produces `Hello Mary` as the output in variable c. This feature can be used along with number-to-string conversion functions, such as `num2str` and `int2str`, to create text objects containing dynamic values of variables. Such text objects are particularly useful in creating titles for figures and other graphics annotation commands (see Section 6). For example, suppose you want to produce a few variance-study graphs with different values of the sample size n, an integer variable. Producing the title of the graph with the command

`title(['Variance study with sample size n = ',int2str(n)])`

writes titles with the current value of n printed in the title.

There are several built-in functions for character string manipulation:

`char`	creates character arrays using automatic padding
	also, converts ASCII numeric values to character arrays
`abs`	converts characters to their ASCII numeric values,
`blanks(n)`	creates n blank characters,
`deblank`	removes the trailing blanks from a string,
`eval`	executes the string as a command (see description below),
`findstr`	finds the specified substring in a given string,
`int2str`	converts integers to strings (see example above),
`ischar`	true (= 1) for character arrays,

isletter	true ($= 1$) for alphabetical characters,
isstring	true ($= 1$) if the argument is a string,
lower	converts any upper case letters in the string to lower case
mat2str	converts a matrix to a string,
num2str	converts numbers to strings (similar to int2str),
strcmp	compares two strings, returns 1 if same,
strncmp	compares the first n characters in given strings,
strcat	concatenates strings horizontally ignoring trailing blanks,
strvcat	concatenates strings vertically ignoring empty strings,
upper	converts any lower case letters in the string to upper case.

The functions **char** and **strvcat** seem to do the same thing — create string arrays by putting each argument string in a row. So, what is the difference? **char** does not ignore empty strings but **strvcat** does. Try the commands: char('top','','bottom'), and strvcat('top','','bottom') to see the difference.

The eval function

MATLAB provides a powerful function **eval** to *evaluate* text strings and execute them if they contain valid MATLAB commands. For example, the command

$$\texttt{eval('x = 5*sin(pi/3)')}$$

computes $x = 5 \sin(\pi/3)$ and is equivalent to typing x = 5*sin(pi/3) on the command prompt.

There are many uses of the **eval** function. One use of this function is in creating or loading sequentially numbered data files. For example, you can use **eval** to run a set of commands ten times while you take a two-hour lunch break. The following script runs the set of commands ten times and saves the output variables in ten different files.

```
% Example of use of EVAL function
% A script file that lets you go out for lunch while MATLAB slogs
%----------------------
t = [0:0.1:1000];                       % t has 10001 elements
for k = 1:10
    outputfile = ['result',int2str(k)];  % see explanation below
    % write commands to run your function here
    theta = k*pi*t;
    x = sin(theta);                     % compute x
    y = cos(theta);                     % compute y
    z = x.*y.^2;                        % compute z
    % now save variables x, y, and z in a Mat-file
    eval(['save ',outputfile,' x y z'])  % see explanation below
end
```

The commands used above are a little subtle. So read them carefully. In particular, note that

- The first command, `outputfile = ...`, creates a name by combining the counter of the `for`-loop with the string `result`, thus producing the names `result1`, `result2`, `result3`, etc. as the loop counter `k` takes the values 1, 2, 3, etc.
- The last command `eval(...)` has *one* input argument—a long string that is made up of three strings: `'save '`, `outputfile`, and `' x y z'`. Note that while `save` and `x, y, z` are enclosed within quotes to make them character strings, the variable `outputfile` is not enclosed within quotes because it is already a string.
- The square brackets inside the `eval` command are used here to produce a single string by concatenating three individual strings. The brackets are not a part of the `eval` command. Note the blank spaces in the strings `'save␣'` and `'␣x␣y␣z'`. These spaces are essential. Can you see why? [Hint: Try to produce the string with and without the spaces.]

Similar to `eval` there is a function `feval` for evaluating functions. See page 93 for a discussion on `feval`.

3.3 Creating and Using *Inline* Functions

A mathematical function, such as $F(x)$ or $F(x, y)$, usually requires just the values of the independent variables for computing the value of the function. We frequently need to evaluate such functions in scientific calculations. One way to evaluate such functions is by programming them in *Function-files* (see Section 4.2 on page 88 for details). However, there is a quicker way of programming functions if they are not too complicated. This is done by defining *inline* functions—functions that are created on the command line. You define these functions using the built-in function `inline`.

The syntax for creating an inline-function is particularly simple:

$$F = \texttt{inline('}\textit{function formula}\texttt{')}$$

Thus, a function such as $F(x) = x^2 \sin(x)$ can be coded as

$$F = \texttt{inline('x\^2 * sin(x)')}$$

However, note that our inline function can only take a scalar `x` as an input argument. We can modify it easily by changing the arithmetic operator to accept array argument: `F = inline('x.^2.*sin(x)')`. Once the function is created, you can use it as a function independently (e.g., type `F(5)` to evaluate the function at $x = 5$), or in the input argument of other functions that can evaluate it. See Fig. 3.5 for more examples and usage of inline-functions. There are several examples of these functions in Section 3.6 on *Plotting Simple Graphs*.

```
>> Fx = inline('x^2+sin(x)');
>> Fx(pi)

ans =

      9.8696
```
Create a simple inline function
$F(x) = x^2 + \sin(x)$, and compute
its value at $x = \pi$.

```
>> Fx = inline('x.^2+sin(x)');
>> x = [0 pi/4 pi/2 3*pi/4 pi];
>> Fx(x);

ans =

      0      -0.0903    1.4674    4.8445    9.8696
```
Modify $F(x)$ so that x can be an
array (i.e., use array operators).
Evaluate $F(x)$ when x is a vector.

```
>> sin(Fx(0.3))

ans =

      -0.2041
```
You can also use $F(x)$ as an input
argument of another function.

```
>> x0 = fzero(Fx, pi/4)

x0 =

      0.8767
```
You can also use $F(x)$ as input
argument to those functions that
require user defined function. Here
we use `fzero` to find a zero of $F(x)$.
See on-line help on `fzero`.

```
>> Fx(x0)

ans =

      2.2204e-016
```

```
>> Fxy = inline('(x.^2+y.^2)./exp(x.*y)');
>> check_sym = [Fxy(1,2)  Fxy(2,1)]

check_sym =

      0.6767        0.6767
```
Create another inline function
$F(x,y)$ of two variables. Note that
here, $F(x,y)$ is symmetric in x & y.
Check its symmetry by evaluating
$F(x,y)$ at (1,2) and (2,1).

Figure 3.5: Examples of creating and using inline functions.

3.4   Using Built-in Functions and On-line Help

For on-line help type:
`help help`

MATLAB provides hundreds of built-in functions for numerical linear algebra, data analysis, Fourier transforms, data interpolation and curve fitting, root-finding, numerical solution of ordinary differential equations, numerical quadrature, sparse matrix calculations, and general-purpose graphics. There is on-line help for all built-in functions. With so many built-in functions, it is important to know how to look for functions and how to learn to use them. There are several ways to get help:

`help`, **the most direct on-line help:** If you know the exact name of a function, you can get help on it by typing `help` *functionname* on the command line. For example, typing `help help` provides help on the function `help` itself.

`lookfor`, **the keyword search help:** If you are looking for a function, use `lookfor` *keyword* to get a list of functions with the string *keyword* in their description. For example, typing `lookfor 'identity matrix'` lists functions (there are two of them) that create identity matrices.

`helpwin`, **the click and navigate help:** If you want to look around and get a feel for the on-line help by clicking and navigating through what catches your attention, use the window based help, `helpwin`. To activate the help window, type `helpwin` at the command prompt or select **Help Window** from the **Help** menu on the command window menu bar.

`helpdesk`, **the web browser based help:** MATLAB 5.x provides extensive on-line documentation in both HTML and PDF formats. If you like to read on-line documentation and get detailed help by clicking on hyperlinked text, use the web browser based help facility, `helpdesk`. To activate the help window, type `helpdesk` at the command prompt or select **Help Desk** from the **Help** menu on the command window menu bar.

As you work more in MATLAB, you will realize that the on-line help with the command `help` and keyword search with `lookfor` are the easiest and the fastest to get help.

Typing `help` by itself brings out a list of categories (see Fig. 3.6) in which help is organized. You can get help on one of these categories by typing `help` *category*. For example, typing `help elfun` gives a list of elementary math functions with a very brief description of each function. Further help can now be obtained on a function because the exact name of the function is now known.

```
>> help

HELP topics:

matlab\general     -  General purpose commands.
matlab\ops         -  Operators and special characters.
matlab\lang        -  Programming language constructs.
matlab\elmat       -  Elementary matrices and matrix manipulation.
matlab\elfun       -  Elementary math functions.
matlab\specfun     -  Specialized math functions.
matlab\matfun      -  Matrix functions - numerical linear algebra.
matlab\datafun     -  Data analysis and Fourier transforms.
matlab\audio       -  Audio support
matlab\polyfun     -  Interpolation and polynomials.
matlab\funfun      -  Function functions and ODE solvers.
matlab\sparfun     -  Sparse matrices.
matlab\graph2d     -  Two dimensional graphs.
matlab\graph3d     -  Three dimensional graphs.
matlab\specgraph   -  Specialized graphs.
matlab\graphics    -  Handle Graphics.
matlab\uitools     -  Graphical user interface tools.
matlab\strfun      -  Character strings.
matlab\iofun       -  File input/output.
matlab\timefun     -  Time and dates.
matlab\datatypes   -  Data types and structures.
matlab\verctrl     -  Version control
matlab\winfun      -  Windows Operating System Interface Files ...
matlab\demos       -  Examples and demonstrations.
  :                   :

For more help on directory/topic, type "help topic".

>> help lang

Programming language constructs.

Control flow.

if         - Conditionally execute statements.
else       - IF statement condition.
elseif     - IF statement condition.
end        - Terminate scope of FOR, WHILE, SWITCH, TRY and IF ..
for        - Repeat statements a specific number of times.
while      - Repeat statements an indefinite number of times.
break      - Terminate execution of WHILE or FOR loop.
switch     - Switch among several cases based on expression.
case       - SWITCH statement case.

  :          :
```

> `help` by itself lists the names of categories in which the on-line help files are organized.

> `help` *category* lists the functions in that category. Detailed help can then be obtained by typing: `help` *functionname.*

Figure 3.6: MATLAB help facility.

On the other hand, `lookfor` is a friendlier command. It lets you specify a descriptive word about the function for which you need help. The following examples take you through the process of looking for help, getting help on the exact function that serves the purpose, and using the function in the correct way to get the results.

Caution: MATLAB's `help` command is not forgiving of any typos or misspellings, and hence you must know the exact command name.

3.4.1 Example–1: Finding the determinant of a matrix

We have a 10×10 matrix `A` (e.g., `A = rand(10)`) and we want to find its determinant. We do not know if the command for determinant is `det`, `deter`, or `determinant`. We can find the appropriate MATLAB command by searching with the `lookfor determinant` command as shown in Fig. 3.7. Then we can find the exact syntax of the command or function. As is evident from the help on `det` in Fig. 3.7, all we have to do is to type `det(A)` to find the determinant of `A`.

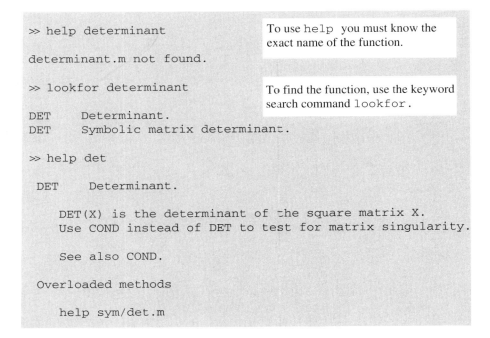

```
>> help determinant                 To use help you must know the
                                     exact name of the function.
determinant.m not found.

>> lookfor determinant              To find the function, use the keyword
                                    search command lookfor.
DET     Determinant.
DET     Symbolic matrix determinant.

>> help det

 DET    Determinant.

    DET(X) is the determinant of the square matrix X.
    Use COND instead of DET to test for matrix singularity.

    See also COND.

 Overloaded methods

    help sym/det.m
```

Figure 3.7: Example of how to find the function that computes the determinant of a matrix.

3.4.2   Example–2: Finding eigenvalues and eigenvectors

Suppose we are interested in finding out the eigenvalues and eigenvectors of a matrix A, but we do not know what functions MATLAB provides for this purpose. Since we do not know the exact name of the required function, our best bet to get help is to try `lookfor eigenvalue`. Figure 3.8 shows MATLAB's response to this command.

```
>> lookfor eigenvalue              lookfor provides keyword search for
                                   help files containing the search string.

ROSSER A classic symmetric eigenvalue test problem.
WILKINSON Wilkinson's eigenvalue test matrix.
BALANCE Diagonal scaling to improve eigenvalue accuracy.
CONDEIG Condition number with respect to eigenvalues.
EIG  Eigenvalues and eigenvectors.
EXPM3 Matrix exponential via eigenvalues and eigenvectors.
POLYEIG Polynomial eigenvalue problem.
QZ   Generalized eigenvalues.
EIGS Find a few eigenvalues and eigenvectors of a matrix...
EIGMOVIE Symmetric eigenvalue movie.
EIGSHOW Graphical demonstration of eigenvalues and ...
   :              :

>> help eig                        Detailed help can then be obtained on
                                   any of the listed files with  help.

 EIG  Eigenvalues and eigenvectors.
   EIG(X) is a vector containing the eigenvalues of a square
   matrix X.

   [V,D] = EIG(X) produces a diagonal matrix D of
   eigenvalues and a full matrix V whose columns are the
   corresponding eigenvectors so that X*V = V*D.

   [V,D] = EIG(X,'nobalance') performs the computation with
   balancing disabled, which sometimes gives more accurate
   results for certain problems with unusual scaling.

   E = EIG(A,B) is a vector containing the generalized
   eigenvalues of square matrices A and B.

   [V,D] = EIG(A,B) produces a diagonal matrix D of general-
   ized eigenvalues and a full matrix V whose columns are
   the corresponding eigenvectors so that A*V = B*V*D.
   :              :
```

Figure 3.8: Example of use of on-line help.

As shown in Fig. 3.8, MATLAB lists all functions that have the string 'eigenvalue' either in their name or in the first line of their description. You can then browse through the list, choose the function that seems closest to your needs, and ask for further help on it. In Fig. 3.8, for example, we seek help on function `eig`. The on-line help on `eig` tells us what this function does and how to use it.

Thus, if we are interested in just the eigenvalues of a matrix A, we type `eig(A)` to get the eigenvalues, but if we want to find both the eigenvalues and the eigenvectors of A, we specify the output list explicitly and type `[eigvec, eigval] = eig(A)` (see Fig. 3.9). The names of the output or the input variables can, of course, be anything we choose. Although it's obvious, we note that the list of the variables (input or output) must be in the same order as specified by the function.

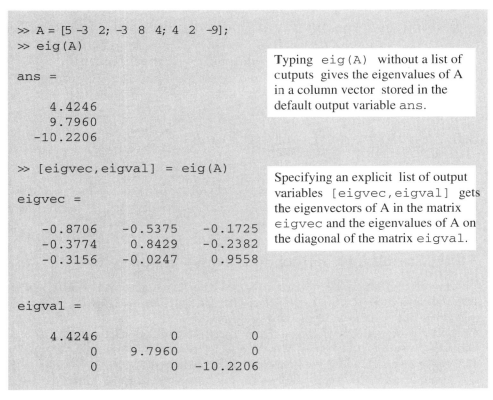

```
>> A = [5 -3  2; -3  8  4; 4  2  -9];
>> eig(A)

ans =

      4.4246
      9.7960
    -10.2206

>> [eigvec,eigval] = eig(A)

eigvec =

    -0.8706   -0.5375   -0.1725
    -0.3774    0.8429   -0.2382
    -0.3156   -0.0247    0.9558


eigval =

      4.4246         0         0
           0    9.7960         0
           0         0  -10.2206
```

Typing `eig(A)` without a list of outputs gives the eigenvalues of A in a column vector stored in the default output variable `ans`.

Specifying an explicit list of output variables `[eigvec,eigval]` gets the eigenvectors of A in the matrix `eigvec` and the eigenvalues of A on the diagonal of the matrix `eigval`.

Figure 3.9: Examples of use of the function `eig` to find eigenvalues and eigenvectors of a matrix.

Some comments on the help facility:

- MATLAB is case-sensitive. All built-in functions in MATLAB use lowercase letters for their names, yet the `help` on any function lists the function in uppercase letters, as is evident from Figs. 3.6 and 3.8. For the first-time user, it may be frustrating to type the command exactly as shown by the help facility, i.e., in uppercase, only to receive an error message from MATLAB.

- On-line help can be obtained by typing help commands on the command line, as we have described in the foregoing section. This method is applicable on all platforms that support MATLAB. In addition, on some computers such as Macintosh and IBM compatibles with Windows, on-line help is also available on the menu bar under **Help**. The menu bar item gives options of opening **Help Window**, a point and click interactive help facility, or **Help Desk**, a web browser based extensive help facility.

- Although `lookfor` is a good command to find help if you are uncertain about the name of the function you want, it has an annoying drawback: it takes only one string as an argument. So typing `lookfor linear equations` causes an error (but `lookfor 'linear equations'` is ok), while both `lookfor linear` and `lookfor equations` do return useful help.

3.5 Saving and Loading Data

For on-line help type:
`help general`

There are many ways of saving and loading data in MATLAB. The most direct way is to use the `save` and `load` commands. You can also save a session or part of a session, including data and commands, using the `diary` command. We will describe `save` and `load` first.

3.5.1 Saving into and loading from the binary Mat-files

The `save` command can be used to save either the entire workspace or a few selected variables in a file called *Mat-file*. Mat-files are written in binary format with full 16 bit precision. It is also possible to write data in Mat-files in 8-digit or 16-digit ASCII format with optional arguments to the `save` command (see the on-line help). Mat-files must always have a '.mat' extension. The data saved in these files can be loaded into the MATLAB workspace by the `load` command. Examples of proper usage of these commands are as follows:

`save tubedata.mat x y`	saves variables `x` and `y` in the file `tubedata.mat`.
`save newdata rx ry rz`	saves variables `rx`, `ry`, and `rz` in the file `newdata.mat`. Here MATLAB automatically supplies the '.mat' extension to the file name.

`save xdata.dat x -ascii`	saves variable `x` in the file `xdata.dat` in 8-digit ASCII format.
`save`	saves the entire workspace in the file `matlab.mat`.
`load tubedata`	loads the variables saved in the file `tubedata.mat`.
`load`	loads the variables saved in the default file `matlab.mat`.

ASCII data files can also be loaded into the MATLAB workspace with the `load` command provided the data file contains only a rectangular matrix of numbers. For more information, see the on-line help on `load`. To read and write ASCII files with specified delimiters (e.g., a tab), use `dlmread` and `dlmwrite`.

You can also use cut-and-paste between the MATLAB command window and other applications (such as Microsoft Excel) to import and export data.

3.5.2 Importing Data Files

MATLAB 6 incorporates a good interface for importing different types of data files, both through a graphical user interface and through the command line. The easiest way to import data from a file is to invoke the *import wizard* by either (i) selecting Import Data from the File menu of the MATLAB window or (ii) typing `uiimport` on the command line, and then following the instructions on the screen. The import wizard does an excellent job of recognizing most file formats, separating data into numeric-data and text-data, and loading the data in MATLAB workspace. Import wizard can load data from several file formats used for text data, spreadsheet data (.xls, .wkl), movie data (.avi), image data (.tiff, .jpeg, etc.), and audio data (.wav, .au). See on-line help on `fileformats` to see a list of readable file formats.

Alternatively, you can also import data from a file using the built-in function `importdata(filename)` from the command line. The *filename* must include the file extension (e.g., .xls for Excel files) so that `importdata` can understand the file format.

For on-line help type:
`help uiimport`
`help iofun`

3.5.3 Recording a session with diary

An entire MATLAB session, or a part of one, can be recorded in a user-editable file, by means of the `diary` command. A file name with any extension can be specified as the output file. For example, typing `diary session1.out` opens a diary file named `session1.out`. Everything in the command window, including user-input, MATLAB output, error messages etc., that follows the diary command is recorded in the file `session1.out`. The recording is terminated by the command `diary off`. The same diary file can be opened later during the same session by typing `diary on`. This will append the subsequent part of the session to the same file `session1.out`. All the figures in this book that show commands typed in the command window and consequent MATLAB output were generated with the diary command. Diary files may be opened and modified with any standard text editor.

The contents of a diary file can be loaded into the MATLAB workspace by converting the diary file into a *script file* (see Section 4.1 for more details on script files) and then executing the script file in MATLAB. To do this, we must first edit the diary file to remove all unwanted lines (for example, MATLAB error messages) and output generated by commands, and then we must rename the file with a '.m' extension in order to make it a script file. Now, we can execute the file by typing the name of the file (without the '.m' extension) at the command prompt. This will execute all the commands in the diary file. Thus, it is also possible to load the values of a variable that were written explicitly in a diary file. If the variables were from an array, we would need to enclose them in square brackets, and assign this to a variable. Values loaded this way will have only the precision explicitly written in the diary file. The number of digits of a variable recorded in the diary depends on the `format` in use during the session (for a description of `format`, see Section 1.6.3). Therefore, we do not recommend this method of saving and loading MATLAB-generated-data. The diary is a good way of recording a session to be included in reports or to show someone how a particular set of calculations was done.

As an alternative to the *diary* facility, you can scroll through the command window and copy (the *cut-and-paste* method) the parts you want to a text file. Another possibility is, if you are a PC user and you have Microsoft Word (MS Office 95 or 97), you can create a fancy **Notebook** —a Word document with embedded MATLAB commands and output, including graphics. To do this you must have a version of MATLAB that supports the *Notebook* facility. Be warned, however, that this facility has not been very reliable in the past because of cross compatibility problems between the various versions of MATLAB and Word.

3.6   Plotting simple graphs

We close this section on interactive computation with an example of how to plot a simple graph and save it as an Encapsulated PostScript file.

As mentioned in the introduction, the plots in MATLAB appear in the graphics window. MATLAB provides very good facilities for both 2-D and 3-D graphics. The commands to produce simple plots are surprisingly simple. For complicated graphics and special effects there are a lot of built-in functions that enable the user to manipulate the graphics window in many ways. Unfortunately, the more control you want the more complicated it gets. We describe the graphics facility in more detail in Chapter 6.

The most direct command to produce a graph in 2-D is the `plot` command. If a variable `ydata` has n values corresponding to n values of variable `xdata`, then `plot(xdata,ydata)` produces a plot with `xdata` on the horizontal axis and `ydata` on the vertical axis. To produce overlay plots, you can specify any number of pairs of vectors as the argument of the `plot` command. We discuss more of this, and much more on other aspects of plotting, in Chapter 6. Figure 3.10 shows an example of plotting a simple graph of $f(t) = e^{t/10}\sin(t),\ \ 0 \le t \le 20$. This function could also be plotted using `fplot`, a command for plotting functions of a single variable. The most

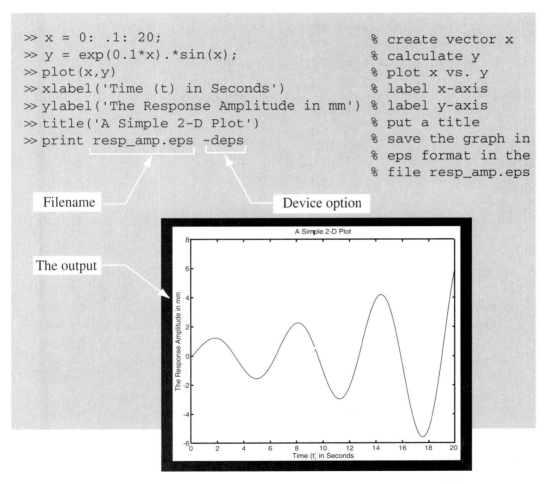

```
>> x = 0: .1: 20;                                  % create vector x
>> y = exp(0.1*x).*sin(x);                         % calculate y
>> plot(x,y)                                        % plot x vs. y
>> xlabel('Time (t) in Seconds')                    % label x-axis
>> ylabel('The Response Amplitude in mm')  % label y-axis
>> title('A Simple 2-D Plot')                       % put a title
>> print resp_amp.eps -deps                         % save the graph in
                                                    % eps format in the
                                                    % file resp_amp.eps
```

Filename   Device option

The output

Figure 3.10: Example of a simple 2-D plot of function $f(t) = e^{t/10}\sin(t)$.

important thing to remember about the `plot` command is that the vector inputs for the x-axis data and the y-axis data must be of the same length. Thus, in the command `plot(x1,y1,x2,y2)`, $x1$, and $y1$ must have the same length, $x2$, and $y2$ must have the same length, while $x1$, and $x2$ may have different lengths.

There are several other plotting functions which are easy to use and quick to produce plots if you are plotting a function:

fplot: It takes the function of a single variable and the limits of the axes as the input and produces a plot of the function. The simplest syntax (there is more to it, see the on-line help) is

fplot('function',[xmin xmax])

Example: The following commands plot $f(x) = e^{-x/10}\sin(x)$ for x between 0 and 20 and produce the plot shown in Fig. 3.11.

```
fplot('exp(-.1*x).*sin(x)',[0, 20])
xlabel('x'),ylabel('f(x) = e^{x/10} sin(x)')
title('A function plotted with fplot')
```

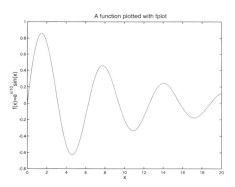

Figure 3.11: Example of plotting a function with `fplot`.

In MATLAB 6 there is a suite of easy function plotters (their names begin with the prefix **ez**) that produce 2-D and 3-D plots and contour plots. These functions require the user to specify the function to be plotted as an input argument. The function to be plotted can be specified as a character string or as an inline-function. There is a default domain for each **ez**-*plotter* but you have the option of overwriting the default domain. These plotting functions are really easy to use. Take a quick look at the following examples.

ezplot: It takes the function of a single variable and the limits of the axes as the input and produces a plot of the function. The syntax is

ezplot('function',[xmin, xmax])

where specification of the domain, $x_{min} < x < x_{max}$ is optional. The default domain is $-2\pi < x < 2\pi$.

Example: The following commands plot $f(x) = e^{-x/10}\cos(x)$ for x between 0 and 20 and produce the plot shown in Fig. 3.12.

```
ezplot('exp(-.1*x).*cos(x)', [0, 20])
```

Alternatively, F = 'exp(-.1*x).*cos(x)';   ezplot(F,[0,20])
or, F = inline('exp(-.1*x).*cos(x)');   ezplot(F,[0,20])
will produce the same result. You can also use `ezplot` with implicit functions (i.e., you do not have an explicit formula for the function as $y = F(x)$, but you have an expression like $x^2 y + \sin(xy) = 0$, or $F(x, y) = 0$). In addition, `ezplot` can also plot parametric curves given by $x(t)$ and $y(t)$ where $t_{min} < t < t_{max}$. See on-line help on `ezplot`.

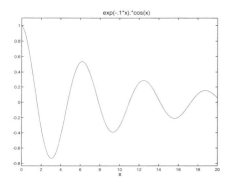

Figure 3.12: Example of plotting a function with `ezplot`.

ezpolar: This is the polar version of `ezplot`. It takes $r(\theta)$ as the input argument. The default domain is $0 < \theta < 2\pi$.
   Example: The following commands plot $r(\theta) = 1 + 2\sin^2(2\theta)$ for $0 < \theta < 2\pi$ and produce the plot shown in Fig. 3.13.

```
r = inline('1+2*(sin(2*t)).^2');   ezplot(r)
```

ezplot3: takes $x(t)$, $y(t)$, and $z(t)$ as input arguments to plot 3-D parametric curves. The default domain is $0 < t < 2\pi$.
   Example: The following commands plot $x(t) = t\cos(3\pi t)$, $y(t) = t\sin(3\pi t)$, and $z(t) = t$ over the default domain. The plot produced is shown in Fig. 3.14.

```
x = 't.*cos(3*pi*t)';   y = 't.*sin(3*pi*t)';   z = 't';
ezplot3(x,y,z)
```

Note that we can also specify x, y, and z as inline-functions (see the next example).

ezcontour: Contour plots are a breeze with this function. Just specify $Z = F(x, y)$ as the input argument and specify the domain for x and y if desired (other than the default $-2\pi < x < 2\pi$, and $-2\pi < y < 2\pi$).
   Example: Let us plot the contours of $Z = \cos x \cos y \, \exp(-\sqrt{(x^2 + y^2)/4})$ over the default domain.

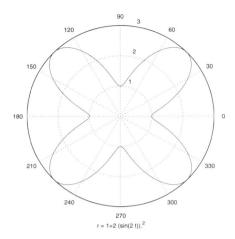

Figure 3.13: Polar plot of $r(\theta) = 1 + 2\sin^2(2\theta)$ with `ezpolar`.

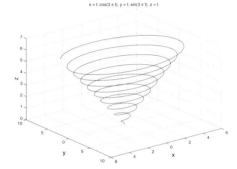

Figure 3.14: Plot of a 3-D parametric space curve with `ezplot3`.

```
Z = inline('cos(x).*cos(y).*exp(-sqrt((x.^2+y.^2)/4))');
ezcontour(Z)
```

The plot produced is shown in Fig. 3.15(a).

ezcontourf: This is just a variant of `ezcontour`. It produces filled contours. The syntax is the same as for `ezcontour`. As an example, let us
take the same function as in `ezcontour` and change the default domain
to $-5 < x < 5$ and $-5 < y < 5$: `ezcontour(Z, [-5,5])`. The plot
produced is shown in Fig. 3.15(b). For different ranges of x and y,
the domain can be specified by a 1-by-4 vector, `[xmin, xmax, ymin,
ymax]`.

ezsurf: To produce stunning surface plots in 3-D, all you need to do is
specify the function $Z = F(x, y)$ as an input argument of `ezsurf` and
specify the domain of the function if you need to. *Example:* Here is an
example of the surface plot for $Z = -5/(1 + x^2 + y^2)$ over the domain

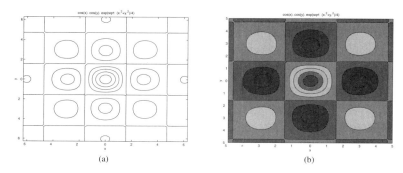

Figure 3.15: Contour plots of $Z = \cos x \cos y \ \exp(-\sqrt{(x^2 + y^2)/4})$ with (a) `ezcontour`, and (b) `ezcontourf`.

$|x| < 3$ and $|y| < 3$.

```
Z = inline('-5./(1+x.^2+y.^2)');
ezsurf(Z,[-3,3,-3,3])
```

The plot produced is similar to the one shown in Fig. 3.16 with no contours.

ezsurfc: An even prettier looking variant of `ezsurf` is `ezsurfc` which combines the surface plot with its contour plot. See Fig. 3.16 that is produced with the command `ezsurfc(Z,[-3,3])` for the same function Z as specified in the `ezsurf` example above.

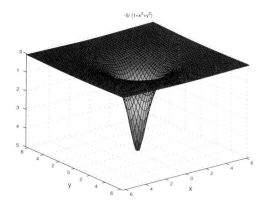

Figure 3.16: 3-D surface plot of $Z = -5/(1 + x^2 + y^2)$ with `ezsurfc`.

There are two more functions in the ez-stable, `ezmesh` and `ezmeshc`, that are used for 3-D plotting. These functions work exactly like the surface plotting functions except that they create the 3-D graphs with wireframe meshes rather than with surface patches.

EXERCISES

1. **Entering matrices:**   Enter the following three matrices.

$$A = \begin{bmatrix} 2 & 6 \\ 3 & 9 \end{bmatrix}, \qquad B = \begin{bmatrix} 1 & 2 \\ 3 & 4 \end{bmatrix}, \qquad C = \begin{bmatrix} -5 & 5 \\ 5 & 3 \end{bmatrix}.$$

2. **Check some linear algebra rules:**

   - **Is matrix addition commutative?** Compute A + B and then B + A. Are the results the same?
   - **Is matrix addition associative?** Compute (A + B) + C and then A + (B + C) in the order prescribed. Are the results the same?
   - **Is multiplication with a scalar distributive?** Compute α(A + B) and αA +αB, taking $\alpha = 5$ and show that the results are the same.
   - **Is multiplication with a matrix distributive?** Compute A*(B+C) and compare with A*B + A*C.
   - **Matrices are different from scalars!** For scalars, $ab = ac$ implies that $b = c$ if $a \neq 0$. Is that true for matrices? Check by computing A*B and A*C for the matrices given above.
     In general, matrix products do not commute either (unlike scalar products). Check if A*B and B*A give different results.

3. **Create matrices with** zeros, eye, **and** ones: Create the following matrices with the help of the matrix generation functions zeros, eye, and ones. See the on-line help on these functions if required (e.g., help eye).

$$D = \begin{bmatrix} 0 & 0 & 0 \\ 0 & 0 & 0 \end{bmatrix}, \qquad E = \begin{bmatrix} 5 & 0 & 0 \\ 0 & 5 & 0 \\ 0 & 0 & 5 \end{bmatrix}, \qquad F = \begin{bmatrix} 3 & 3 \\ 3 & 3 \end{bmatrix}.$$

4. **Create a big matrix with submatrices:** The following matrix G is created by putting matrices A, B, and C given above, on its diagonal. In how many ways can you create this matrix using submatrices A, B, and C (that is, you are not allowed to enter the non-zero numbers explicitly)?

$$G = \begin{bmatrix} 2 & 6 & 0 & 0 & 0 & 0 \\ 3 & 9 & 0 & 0 & 0 & 0 \\ 0 & 0 & 1 & 2 & 0 & 0 \\ 0 & 0 & 3 & 4 & 0 & 0 \\ 0 & 0 & 0 & 0 & -5 & 5 \\ 0 & 0 & 0 & 0 & 5 & 3 \end{bmatrix},$$

5. **Manipulate a matrix:** Do the following operations on matrix G created above in (4).

   - Delete the last row and last column of the matrix.
   - Extract the first 4×4 submatrix from G.

- Repalce G(5,5) with 4.
- What do you get if you type `G(13)` and hit return? Can you explain how MATLAB got that answer?
- What happens if you type `G(12,1) = 1` and hit return?

6. **See the structure of a matrix:** Create a 20×20 matrix with the command `A = ones(20)`. Now replace the 10×10 submatrix between rows 6:15 and columns 6:15 with zeros. See the structure of the matrix (in terms of nonzero entries with the command `spy(A)`. Set the 5×5 submatrices in the top right corner and bottom left corner to zeros and see the structure again.

7. **Create a symmetric matrix:** Create an upper triangular matrix with the following command:

$$A = \text{diag}(1:6) + \text{diag}(7:11,1) + \text{diag}(12:15,2).$$

Make sure you understand how this command works (see the on-line help on `diag` if required). Now use the upper off-diagonal terms of A to make A a symmetric matrix with the following command:

$$A = A + \text{triu}(A,1)'.$$

This is a somewhat *loaded* command. It takes the upper triangular part of A above the main diagonal, flips it (transposes), and adds to the original matrix A, thus creating a symmetric matrix A. See the on-line help on `triu`.

8. **Do some cool operations:** Create a 10×10 random matrix with the command `A = rand(10)`. Now do the following operations.

- Multiply all elements by 100 and then round off all elements of the matrix to integers with the command `A = fix(A)`.
- Replace all elements of $A < 10$ with zeros.
- Replace all elements of $A > 90$ with infinity (`inf`).
- Extract all $30 \leq a_{ij} \leq 50$ in a vector b, that is, find all elements of A that are between 30 and 50 and put them in a vector b.

9. **How about some fun with plotting?**

- Plot the parametric curve $x(t) = t$, $y(t) = e^{-t/2} \sin t$ for $0 < t < \pi/2$ using `ezplot`.
- Plot the cardioid $r(\theta) = 1 + \cos\theta$ for $0 < \theta < 2\pi$ using `ezpolar`.
- Plot the contours of $x^2 + \sin(xy) + y^2 + z^2 = 0$ using `ezcontour` over the domain $-\pi/2 < x < \pi/2$, $-\pi/2 < y < \pi/2$.
- Create a surface plot along with contours of the function $H(x,y) = \frac{x^2}{2} + (1 - \cos y)$ for $-\pi < x < \pi$, $-2 < y < 2$.

4. Programming in MATLAB: Scripts and Functions

For on-line help type:
`help lang`

A distinguishing features of MATLAB is its *ease* of extendability, through user-written programs. MATLAB provides its own language, which incorporates many features from C. In some regards, it is a higher-level language than most common programming languages, such as Pascal, Fortran, and C, meaning that you will spend less time worrying about formalisms and syntax. For the most part, MATLAB's language feels somewhat natural.

In MATLAB you write your programs in *M-files*. M-files are ordinary ASCII text files written in MATLAB's language. They are called M-files because they must have a '.m' at the end of their name (like `myfunction.m`). M-files can be created using any editor or word processing application.

There are two types of M-files—*script* files and *function* files. Now we discuss their purpose and syntax.

4.1 Script Files

A script file is an 'M-file' with a set of valid MATLAB commands in it. A script file is executed by typing the name of the file (without the '.m' extension) on the command line. It is equivalent to typing all the commands stored in the script file, one by one, at the MATLAB prompt. Naturally, script files work on global variables, that is, variables currently present in the workspace. Results obtained from executing script files are left in the workspace.

A script file may contain any number of commands, including those that call built-in functions or functions written by you. Script files are useful when you have to repeat a set of commands several times. Here is an example.

Example of a Script File: Let us write a script file to solve the following system of linear equations:[1]

$$
\begin{bmatrix}
5 & 2r & r \\
3 & 6 & 2r-1 \\
2 & r-1 & 3r
\end{bmatrix}
\begin{Bmatrix}
x_1 \\
x_2 \\
x_3
\end{Bmatrix}
=
\begin{Bmatrix}
2 \\
3 \\
5
\end{Bmatrix}
\tag{4.1}
$$

or $\mathbf{Ax} = \mathbf{b}$. Clearly, $\mathbf{A}$ depends on the parameter r. We want to find the solution of the equation for various values of the parameter r. We also want to find, say, the determinant of matrix $\mathbf{A}$ in each case. Let us write a set of MATLAB commands that do the job, and store these commands in a file called 'solvex.m'. How you create this file, write the commands in it, and save the file, depends on the computer you are using. In any case, you are creating a file called `solvex.m`, which will be saved on some disk drive in some directory (or folder).

```
%----------- This is the script file 'solvex.m' -----------
% It solves equation (4.1) for x and also calculates det(A).

A = [5 2*r r; 3 6 2*r-1; 2 r-1 3*r];   % create matrix A
b = [2;3;5];                           % create vector b
det_A = det(A)                         % find the determinant
x = A\b                                % find x
```

In this example, we have not put a semicolon at the end of the last two commands. Therefore, the results of these commands will be displayed on the screen when we execute the script file. The results will be stored in variables det_A and x, and these will be left in the workspace.

Let us now execute the script in MATLAB.

---

[1]If you are not familiar with matrix equations, see Section 5.1.1 on page 117.

```
>> r = 1;              % specify a value of r
>> solvex             % execute the script file solvex.m

det_A =

      64
x =
    -0.0312
     0.2344
     1.6875
```

> This is the output. The values of the variables det_A and x appear on the screen because there is no semi-colon at the end of the corresponding lines in the script file.

You may notice that the value of **r** is assigned outside the script file and yet **solvex** picks up this value and computes **A**. This is because all the variables in the MATLAB workspace are available for use within the script files, and all the variables in a script file are left in the MATLAB workspace. Even though the output of **solvex** is only **det_A** and **x**, **A** and **b** are also there in your workspace, and you can see them after execution of **solvex** by typing their names at the MATLAB prompt. So, if you want to do a big set of computations but in the end you want only a few outputs, a script file is not the right choice. What you need in this case is a *function file*.

Caution:

- *NEVER name a script file the same as the name of a variable it computes.* When MATLAB looks for a name, it first searches the list of variables in the workspace. If a variable of the same name as the script file exists, MATLAB will never be able to access the script file. Thus, if you execute a script file **xdot.m** that computes a variable **xdot**, then after the first execution the variable **xdot** exists in the MATLAB workspace. Now if you change something in the script file and execute it again, all you get is the old value of **xdot**! Your changed script file is not executed because it cannot be accessed as long as the variable **xdot** exists in the workspace.

- The name of a script file must begin with a letter. The rest of the characters may include digits and the underscore character. You may give long names but MATLAB will take only the first 19 characters. You may not use any periods in the name other than the last one in '.m'. Thus, names such as **project_23C.m**, **cee213_hw5_1.m**, and **MyHeartThrobsProfile.m** are fine but **project.23C.m** and **cee213_hw5.1.m** are not valid names.

- Be careful with variable names while working with script files, because all variables generated by a script file are left in the workspace, unless you clear them. Avoid name clashes with built-in functions. It is a good idea to first check if a function or script file of the proposed name already exists. You can do this with the command `exist('name')`, which returns zero if nothing with name *name* exists.

4.2   Function Files

A function file is also an m-file, like a script file, except that the variables in a function file are all local. Function files are like programs or subroutines in Fortran, procedures in Pascal, and functions in C. Once you get to know MATLAB well, this is where you are likely to spend most of your time — writing and refining your own function files.

A function file begins with a function definition line, which has a well-defined list of inputs and outputs. Without this line, the file becomes a script file. The syntax of the function definition line is as follows:

```
function [output variables] = function_name(input variables);
```

where the *function_name* must be the same as the *filename* (without the '.m' extension) in which the function is written. For example, if the name of the function is `projectile` it must be written and saved in a file with the name `projectile.m`. The function definition line may look slightly different depending on whether there is no output, a single output, or multiple output.

Examples:

Function Definition Line	File Name
function [rho,H,F] = motion(x,y,t);	motion.m
function [theta] = angleTH(x,y);	angleTH.m
function theta = THETA(x,y,z);	THETA.m
function [] = circle(r);	circle.m
function circle(r);	circle.m

Caution: The first word in the function definition line, `function`, *must be typed in lower case*. A common mistake is to type it as `Function`.

Anatomy of a function file

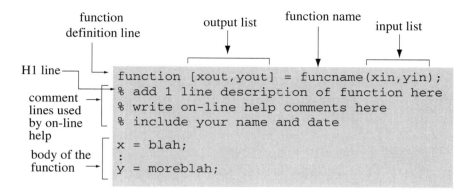

- Comment lines start with a '%' sign and may be put anywhere. Anything after a % in a line is ignored by MATLAB as a non-executable statement.

- All comment lines immediately following the function definition line are displayed by MATLAB if **help** is sought on the function. The very first comment line immediately following the definition line is called the 'H1' line. An H1 line, if present, is automatically cataloged in the *Contents.m* file of the directory in which the file resides. This allows the line to be referenced by the **lookfor** command. A word of caution: any blanks before the % sign in the first comment line disqualify it from becoming an H1 line. Welcome to the idiosyncrasies of MATLAB!

- A single-output variable is not required to be enclosed in square brackets in the function definition line, but multiple output variables must be enclosed within []. When there is no output variable present, the brackets as well as the equal sign may be omitted (see examples above).

- Input variable names given in the function definition line are local to the function, so other variable names or values can be used in the function call. The name of another function can also be passed as an input variable. No special treatment is required for the function names as input variables in the function definition line. However, when the function is executed, the name of the input function must be passed as a character string, i.e., enclosed within two single right quotes (see example in the next section).

4.2.1 Executing a function

There are two ways a function can be executed, whether it is built-in or user-written:

1. **With explicit output:** This is the full syntax of calling a function. Both the output and the input list are specified in the call. For example, if the function definition line of a function reads

     `function [rho,H,F] = motion(x,y,t);`

   then all the following commands represent legal call (execution) statements:

   - `[r,angmom,force] = motion(xt,yt,time);` The input variables `xt`, `yt`, and `time` must be defined before executing this command.
   - `[r,h,f] = motion(rx,ry,[0:100]);` The input variables `rx` and `ry` must be defined beforehand, the third input variable `t` is specified in the call statement.
   - `[r,h,f] = motion(2,3.5,0.001);` All input values are specified in the call statement.
   - `[radius,h] = motion(rx,ry);` Call with partial list of input and output. The third input variable must be assigned a default value inside the function if it is required in calculations. The output corresponds to the first two elements of the output list of `motion`.

2. **Without any output:** The output list can be omitted entirely if the computed quantities are not of any interest. This might be the case when the function displays the desired result graphically. To execute the function this way, just type the name of the function with the input list. For example, `motion(xt,yt,time);` will execute the function `motion` without generating any explicit output, provided that `xt`, `yt`, and `time` are defined, of course. If the semicolon at the end of the call statement is omitted, the first output variable in the output-list of the function is displayed in the default variable `ans`.

A function can be written to accept a partial list of inputs if some default values of the other unspecified inputs is defined inside the function. This kind of input list manipulation can be done with the built-in function `nargin`, which stands for **number-of-arguments-in**. Similarly, the list of output can be manipulated with the built-in function `nargout`. See the on-line help on `nargin` and `nargout`. For an example of how to use them, look at the function `fplot` by typing `type fplot.m`.

Example of a Simple Function File:

Let us write a function file to solve the same system of linear equations that we solved in Section 4.1 using a script file. This time, we will make r an input to the function and det_A and x will be the output. Let us call this function **solvexf**. *As a rule, it must be saved in a file called* **solvexf.m**.

```
function [det_A, x] = solvexf(r);
% SOLVEXF solves a 3X3 matrix equation with parameter r
% This is the function file 'solvexf.m'
% To call this function, type:
%   [det_A,x] = solvexf(r);
% r is the input and det_A and x are output.
%------------------------------------------------

A = [5 2*r r; 3 6 2*r-1; 2 r-1 3*r]; % create matrix A
b = [2;3;5];                         % create vector b
det_A = det(A);                      % find the determinant
x = A\b;                             % find x.
```

Now r, x and *det_A* are all local variables. Therefore, any other variable names may be used in their places in the function call statement. Let us execute this function in MATLAB .

```
>> [detA, y] = solvexf(1); % take r=1 and execute solvexf.m

>> detA                    % display the value of detA

ans =
      64

>> y                       % display the value of y

ans =
    -0.0312                Values of detA and y will be
     0.2344                automatically displaced if the semi-
     1.6875                colon at the end of the function
                           execution command is omitted.
```

After execution of a function, the only variables left in the workspace by the function will be the variables in the output list. This gives us more control over input and output than we can get with script files. We can also include error checks and messages inside the function. For example, we could modify the function above to check if matrix A is empty or not and display an appropriate message before solving the system, by changing the last line to:

```
if isempty(A)              % if matrix A is empty
    disp('Matrix A is empty');
else                       % if A is not empty
    x = A\b;               % find x
end                        % end of if statement.
```

For a description of `if-elseif-else` branching and other control flow commands, see Section 4.3.4 on page 98.

4.2.2   More on functions

By now the difference between scripts and functions should be clear to you. The variables inside a function are local and are erased after execution of the function. But the variables inside a script file are left in the MATLAB workspace after execution of the script. Functions can have arguments, script files do not. What about functions inside another function? Are they local? How are they executed? Can a function be passed as an input variable to another function? Now we address these questions.

Executing a function inside another function

Usually, it is a straightforward process, so much that you do not have to pay any special attention to it. In the function `solvexf` above, we used a built-in function, `det`, to calculate the determinant of **A**. We used the function just the way we would use it at the MATLAB prompt or in a script file. This is true for all functions, built-in or user-written. The story is different only when you want the function name to be *dynamic*, that is, if the function to be executed inside may be different with different executions of the calling function[2]. In such cases, the actual name of the function is passed to the calling function through the input list and a dummy name is used inside the calling function. The mechanism of passing the function name and evaluating it inside the calling function is quite different from that for a variable. We explain it below.

A function in the input list:

When a function needs to be passed in the input list of another function, the name of the function to be passed must appear as a character string in the input list. For example, the built-in function `fzero` finds a zero of a user-supplied function of a single variable. The call syntax of the function is `fzero(f,x)` where `f` is the name of the function or a *function handle*, and `x` is an initial guess. There are several ways in which we can code the function 'f' and pass it in the input list of `fzero`. We describe a few common ways of doing so below on an example function

$$f(r) = r^3 - 32.5r^2 + (r - 22)r + 100.$$

---

[2]A function that uses another function inside its body is called a *calling function*. For example, `solvexf` in the example above is a calling function for the function `det`.

Use inline function: We can make $f(r)$ to be an inline function (see Section 3.3 on page 67 for a detailed discussion) and pass its name to `fzero` as follows.

```
fun = r.^3 - 32.5*r.^2 + (r-22).*r + 100;
r0 = fzero(fun,5);
```

Use function file: We can code $f(r)$ in a function file called `funr.m` as follows.

```
function f = funr(r);
%FUNR evaluates function f = r^3 -32.5 r^2+(r-22)r+100.
f = r.^3 - 32.5*r.^2 + (r-22).*r + 100;
```

Now we may call `fzero` with the statement — fzero('funr',5). Note the single quotes around the name of the input function `funr`.

Use function handle: A *function handle* is a convenient function identifier (a variable) created by the user. It is created with the @ symbol. We can use it for any function—inline, built-in, or user written M-file function. As examples, let us do the same thing with `fun` and `funr` as we did above but with function handles.

For on-line help type:
`help @`

```
f1 = @fun;          % create function handle f1 for 'fun'
r1 = fzero(f1,5);   % use function handle

f2 = @funr          % create function handle f2 for 'funr'
r2 = fzero(f2,5);   % use function handle
```

Of course, we could compress two statements for each call in a single statement: `r1 = fzero(@fun,5)` or `r2 = fzero(@funr,5)`.

Evaluating a function with `feval`

The function `feval` evaluates a function whose name is specified as a string at the given list of input variables. For example [y, z] = feval ('Hfunction', x, t); evaluates the function `Hfunction` on the input variables `x` and `t` and returns the output in `y` and `z`. It is equivalent to typing [y, z] = Hfunction (x,t). So why would you ever evaluate a function using `feval` when you can evaluate the function directly? The most common use is when you want to evaluate functions with different names but the same input list. Consider the following situation. You want to evaluate any given function of x and y at the origin $x = 0$, $y = 0$. You can write a script file with one command in it:

```
value = feval ('funxy', 0, 0);       .
```

This command is the same as writing, `value = funxy(0,0)`. Now suppose Harry has a function $z(x,y) = \sin xy + xy^2$, programmed as:

```
function  z = harrysf(x,y)
% function to evaluate z(x,y)
z = sin(x*y) + x*y^2;
```

and Kelly has a function $h(x, y) = 20xy - 3y^3 - 2x^3 + 10$, programmed as

```
function  h = kellysf(x,y)
% function to evaluate h(x,y)
h = 20*x*y - 3*y^3 - 2*x^3 + 10;
```

Both functions can be evaluated with your script file by changing the name `funxy` to `harrysf` and `kellysf`, respectively. The point here is that the command in your script file takes *dynamic* filenames.

The use of feval becomes essential[3] when a function is passed as an input variable to another function. In such cases, the function passed as the input variable must be evaluated using `feval` inside the calling function. For example, the ODE (ordinary differential equation) solvers `ode23` and `ode45` take user-defined functions as inputs in which the user specifies the differential equation. Inside `ode23` and `ode45`, the user-defined function is evaluated at the current time t and the current value of x to compute the derivative $\dot{x}$ using `feval`. `ode23` and `ode45` are M-files, which you can copy and edit. Make a printout of one of them and see how it uses `feval`.

Writing good functions

Writing functions in MATLAB is easier than writing functions in most standard programming languages, or, for that matter, in most of the software packages that support their own programming environment. However, writing efficient and elegant functions is an art that comes only through experience. For beginners, keeping the following points in mind helps.

- **Pseudo-code:** Before you begin writing a function, write a *pseudocode*. It is essentially the entire function in plain English. Think about the logical structure and the sequence of computations, define the input and output variables, and write the function in plain words. Then begin the translation into MATLAB language.

- **Readability:** Select a sensible name for the function and the variables inside it. Write enough comments in the body of the function. Design and write helpful comments for on-line help (the comment lines in the beginning of the function). Make sure you include the syntax of how to use the function.

---

[3]Unless you are comfortable with `eval` (see page 66) and can replace `feval(fun,x,y)` with `eval([fun,'(x,y)'])` where `fun` is a character string containing the name of the desired function.

- **Modularity:** Keep your functions *modular*, that is, break big computations into smaller chunks and write separate functions for them. Keep your functions small in length.

- **Robustness:** Provide checks for errors and exit with helpful error messages.

- **Expandability:** Leave room for growth. For example, if you are writing a function with scalar variables, but you think you may use vector variables later, write the function with this in mind. Avoid hardcoding actual numbers.

4.2.3   Subfunctions

MATLAB 5 allows several functions to be written in a single file. While this facility is useful from file organization point of view, it comes with severe limitations. All functions written below the first function in the file are treated as subfunctions and they are NOT available to the outside world. This means that while the first function can access all its subfunctions, and the subfunctions written in the same file can also access each other, functions outside this file cannot access these subfunctions.

4.2.4   Compiled (Parsed) functions: *P-Code*

When a function is executed in MATLAB, each command in the function is first interpreted by MATLAB and then translated into the lower level language. This process is called *parsing*. It is not exactly like compiling a function in C or Fortran but, superficially, it is a similar process. MATLAB 5 allows you to save a parsed function for later use. To parse a function called `projectile`, that exists in an M-file called `projectile.m`, type the command

```
pcode projectile
```

This command generates a parsed function that is saved in a file named `projectile.p`. Next time, you call the function projectile, MATLAB directly executes the preparsed function.

For all moderate size functions, saving them as parsed functions does not save as much time during execution as you would think. The MATLAB parser is quick enough to reduce the overhead on parsing to a negligible fraction. The best use of p-codes, perhaps, is in protecting your proprietary rights when you send your functions to other people. Send p-codes, the recipients can execute them but they cannot modify them.

4.2.5   The *Profiler*

To evaluate the performance of functions, MATLAB 5 provides a tool called *profiler*. The profiler keeps track of the time spent on each line of the function (in units of 0.01 second) as the function is executed. The profile report shows how much time was spent on which line. The profiler output can also be seen graphically as a pareto plot.

The profiler is surprisingly easy to use. Let us see a pseudo example. To profile a function called `projectile`, type the following commands:

```
profile on                      % turn the profiler on
[a,b] = projectile(v,theta)     % execute the function projectile
profile report                  % show the profiler report
profile plot                    % show the report as a pareto plot
profile done                    % thank you, bye
```

Note: The profiler can be used only on functions (not scripts) and those functions that exist as M-files. So, you cannot profile the built-in functions that are not provided as M-files.

For on-line help type:
`help lang`

4.3   Language-Specific Features

We have already discussed numerous features of MATLAB's language through many examples in the previous sections. You are advised to pay special attention to proper usage of punctuation marks and different delimiters (page 227), and operators, especially the array operators (a period (.) preceding the arithmetic operators, page 58) and the relational operators (page 59). For control-flow, MATLAB provides `for` and `while` loops, an `if-elseif-else` construct, and a `switch-case-otherwise` construct. All the control-flow statements must terminate with corresponding `end` statements. We now discuss flow control and some other specific features of the language. See the on-line help for more details.

4.3.1   Use of comments to create on-line help

As we have already pointed out in the discussion on function files (page 89), the comment lines at the beginning (before any executable statement) of a script or a function file are used by MATLAB as the on-line help on that file. This automatically creates the on-line help for user-written functions. It is a good idea to copy the function definition line without the word `function` among those first few comment lines so that the execution syntax of the function is displayed by the on-line help. The command `lookfor` looks for the argument string in the first commented line of m-files. Therefore, in keeping with the somewhat confusing convention of MATLAB's built-in functions, you should write the name of the script or function file in uppercase letters followed by a short description with keywords, as the first commented line. Thus the first line following the function definition line in the example function on page 90 reads

```
% SOLVEXF solves a 3x3 matrix equation with parameter r.
```

4.3.2   Continuation

An ellipsis (three consecutive periods, '...') at the end of a line denotes continuation. So, if you type a command that does not fit on a single line, you may split it across two or more lines by using an ellipsis at the end of each but the last line. *Examples:*

```
A = [1 3 3 3; 5 10 -2 -20; 3 5 ...
     10 2; 1 0 0 9];
x = sin(linspace(1,6*pi,100)) .* cos(linspace(1,6*pi,100)) +...
    0.5*ones(1,100);
plot(tube_length,fluid_pressure,':',tube_length,...
     theoretical_pressure,'-')
```

You *may not*, however, use continuation inside a character string. For example, typing

```
logo = 'I am not only the President and CEO of Miracle Hair,...
        but also a client';
```

produces an error. For creating such long strings, break the string into smaller string segments and use concatenation (see Section 3.2.6).

4.3.3   Global variables

It is possible to declare a set of variables to be globally accessible to all or some functions without passing the variables in the input list. This is done with the `global` command. For example, the statement `global x y z` declares the variables `x`, `y`, and `z` to be global. This statement goes before any executable statement in the functions and scripts that need to access the values of the global variables. Be careful with the names of the global variables. It is generally a good idea to name such variables with long strings to avoid any unintended match with other local variables.

   Example: Consider solving the following 1st-order ODE:

$$\dot{x} = kx + c\sin t, \qquad x(0) = 1.0$$

where you are interested in solutions for various values of k and c. Your script file may look like:

```
% scriptfile to solve a 1st-order ode.
ts = [0  20];                   % specify time span = [t_0 t_final]
x0 = 1.0;                       % specify initial condition
[t, x] = ode23 ('ode1',ts,x0); % execute ode23 to solve the ODE.
```

and the function 'ode1' may look like:

```
function xdot = ode1(t,x);
% ODE1: function to compute the derivative xdot
% at given t and x.
% Call syntax: xdot = ode1 (t,x);
% -------------
xdot = k*x + c*sin(t);
```

This, however, won't work. In order for **ode1** to compute **xdot**, the values of k and c must be prescribed. These values could be prescribed inside the function **ode1** but you would have to edit this function each time you change the values of **k** and **c**. An alternative[4] is to prescribe the values in the script file and make them available to the function **ode1** through *global* declaration.

```
% scriptfile to solve 1st-order ode.
global k_value c_value          % declare global variables
k_value = 5;  c_value = 2;      % specify the values of global variables
ts = [0 20];                    % specify time span
x0 = 1.0;                       % specify initial condition
[t, x] = ode23('ode1',ts,x0);  % execute ode23 to solve the ODE.
```

Now you have to modify the function **ode1** so that it can access the global variables:

```
function xdot = ode1(t,x);
% ODE1: function to compute the derivative xdot
% at given t and x.
% Call syntax: xdot = ode1(t,x);
% -------------
global k_value c_value
xdot = k_value*x + c_value*sin(t)
```

Now, if the values of **k_value** and **c_value** are changed in the script file, the new values become available to **ode1** too. Note that the global declaration is only in the script file and the user function file **ode1**, and therefore **k_value** and **c_value** will be available to these files only.

For on-line help type:
help lang

4.3.4   Loops, branches, and control-flow

MATLAB has its own syntax for control-flow statements like *for*-loops, *while*-loops and, of course, *if-elseif-else* branching. In addition, it provides three commands—**break, error,** and **return** to control the execution of scripts and functions. A description of these functions follows.

---

[4]Another alternative is to use **ode23** (of MATLAB 5.x) to pass the variables k and c to **ode1**. See Section 5.5.5 on page 146 for details.

For loops:

A `for` loop is used to repeat a statement or a group of statements for a fixed number of times. Here are two examples:

Example-1:
```
for m=1:100
    num = 1/(m+1)
end
```

Example-2:
```
for n=100:-2:0, k = 1/(exp(n)), end
```

The *counter* in the loop can also be given explicit increment: `for i=m:k:n` to advance the counter `i` by `k` each time (in the second example above `n` goes from 100 to 0 as 100, 98, 96, ..., etc.). You can have nested `for`-loops, that is, `for`-loops within `for`-loops. *Every* `for`, *however, must be matched with an* `end`.

While loops:

A `while` loop is used to execute a statement or a group of statements for an indefinite number of times until the condition specified by `while` is no longer satisfied. For example:

```
% let us find all powers of 2 below 10000
v = 1;   num = 1;  i=1;
while num < 10000
        num = 2^i;
        v = [v; num];
        i = i + 1;
end
```

Once again, a `while` must have a matching `end`.

If-elseif-else statements:

This construction provides a logical branching for computations. Here is an example:

```
i=6; j=21;
if   i > 5
        k = i;
elseif (i>1) & (j==20)
        k = 5*i+j;
else
        k = 1;
end
```

Of course, you can have nested `if` statements, as long as you have matching `end` statements. You can nest all three kinds of loops, in any combination.

Switch-case-otherwise:

This construction (a feature of MATLAB 5.x) provides another logical branching for computations. A flag (any variable) is used as a *switch* and the values of the flag make up the different *cases* for execution. The general syntax is:

```
switch flag
case value1
        block 1 computation
case value2
        block 2 computation
otherwise
        last block computation
end
```

Here, *block 1 computation* is performed if flag = value1, *block 2 computation* is performed if flag = value2, and so on. If the value of flag does not match any case then the *last block computation* is performed. Of course, like all good things in life, the switch must come to an end too.

The switch can be a numeric variable or a string variable. Let us look at a more concrete example using a string variable as the switch:

```
color = input('color = ','s');
switch color
case 'red'
        c = [1 0 0];
case 'green'
        c = [0 1 0];
case 'blue'
        c = [0 0 1];
otherwise
        error('invalid choice of color')
end
```

Break

The command break inside a for or while loop terminates the execution of the loop, even if the condition for execution of the loop is true. *Examples:*

```
1.      for i=1:length(v)
            if u(i) < 0        % check for negative u
               break           % terminate loop execution
            end
            a = a + v(i);      % do something
        end
```

2.
```
        x = exp(sqrt(163));
        while 1
            n = input('Enter max. number of iterations ')
            if n <= 0
                break              % terminate loop execution
            end
            for i=1:n
                x = log(x);   % do something
            end
        end
```

If the loops are nested then **break** terminates only the innermost loop.

Error

The command **error('*message*')** inside a function or a script aborts the execution, displays the error message *message*, and returns the control to the keyboard.

Example:
```
    function c = crossprod(a,b);
    % crossprod(a,b) calculates the cross product axb.
    if nargin~=2        % if not two input arguments
        error('Sorry, need two input vectors')
    end
    if length(a)==2     % begin calculations
        ....
    end
```

Return

The command **return** simply returns the control to the invoking function.

Example:
```
    function animatebar(t0,tf,x0);
    % animatebar animates a bar pendulum.
    :
    disp('Do you want to see the phase portrait?')
    ans = input('Enter 1 if YES, 0 if NO '); % see below for description
    if ans==0               % if the input is 0
        return              % exit function
    else
        plot(x,...)         % show the phase plot
    end
```

4.3.5   Interactive input

The commands—input, keyboard, menu, and pause can be used inside a script or a function file for interactive user input. Their descriptions follow.

For on-line help type:
`help lang`

Input

The command input(*'string'*), used in the previous example, displays the text in *string* on the screen and waits for the user to give keyboard input.

Examples:

- n = input('Largest square matrix size '); prompts the user to input the size of the 'largest square matrix' and saves the input in n.

- more = input('More simulations? (Y/N) ','s'); prompts the user to type Y for YES and N for NO and stores the input as a string in more. Note that the second argument, 's', of the command directs MATLAB to save the user input as a string.

This command can be used to write *user-friendly* interactive programs in MATLAB.

Keyboard

The command keyboard inside a script or a function file returns control to the keyboard at the point where the command occurs. The execution of the function or the script is *not* terminated. The command window prompt '≫ ' changes to 'k≫ ' to show the special status. At this point, you can check variables already computed, change their values, and issue any valid MATLAB commands. The control is returned to the function by typing the word return on the special prompt k≫ and then pressing the return key.

This command is useful for debugging functions. Sometimes, in long computations, you may like to check some intermediate results, plot them and see if the computation is headed in the right direction, and then let the execution continue.

Example:

```
% EXKEYBRD: a script file for example of keyboard command

A = ones(10)              % make a 10x10 matrix of 1's
for i=1:10
    disp(i)               % display the value of i
    A(:,i) = i*A(:,i);    % replace the ith column of a
    if i==5               % when i = 5
        keyboard          % return the control to keyboard
    end
end
```

During the execution of the above script file exkeybrd.m, the control is returned to the keyboard when the value of the counter i reaches 5. The execution of exkeybrd resumes after the control is returned to the script file by typing return on the special prompt k≫ .

Menu

The command `menu('MenuName','option1','option2',..)` creates an
on-screen menu with the *MenuName* and lists the options in the menu. The
user can select any of the options using the mouse or the keyboard depending
on the computer. The implementation of this command on Macs and PCs
creates nice window menu with buttons.
Example:

```
% Plotting a circle
r = input('Enter the desired radius ');
theta = linspace(0,2*pi,100);
r = r*ones(size(theta));     % make r the same size as theta
coord = menu('Circle Plot','Cartesian','Polar');
if coord==1                  % if the first option is selected
                             %- from the menu
    plot(r.*cos(theta),r.*sin(theta))
    axis('square')
else                         % if the second option is selected
                             %- frcm the menu
    polar(theta,r);
end
```

In the above script file, the `menu` command creates a menu with the name
Circle Plot and two options—Cartesian and Polar. The options are internally
numbered. When the user selects one of the options, the corresponding
number is passed on to the variable `coord`. The `if-else` construct following
the `menu` command shows what to do with each option. Try out this script
file.

Pause

The command `pause` temporarily halts the current process. It can be used
with or without an optional argument:

pause   halts the current process and waits for the user to give a
          'go-ahead' signal. Pressing any key resumes the process.
          Example: `for i=1:n, plot(X(:,i),Y(:,i)), pause, end`.
pause(*n*) halts the current process, pauses for *n* seconds,
          and then resumes the process.
          Example: `for i=1:n, plot(X(:,i),Y(:,i)), pause(5), end`
                    pauses for 5 seconds before it plots the next graph.

4.3.6   Recursion

The MATLAB programming language supports recursion, i.e., a function can
call itself during its execution. Thus recursive algorithms can be directly
implemented in MATLAB (what a break for Fortran users!).

4.3.7 Input/output

For on-line help type:
`help iofun`

MATLAB supports many standard C-language file I/O functions for reading and writing formatted binary and text files. The functions supported include:

`fopen`	opens an existing file or creates a new file
`fclose`	closes an open file
`fread`	reads binary data from a file
`fwrite`	writes binary data to a file
`fscanf`	reads formatted data from a file
`fprintf`	writes formatted data to a file
`sscanf`	reads strings in specified format
`sprintf`	writes data in formatted string
`fgets`	reads a line from file discarding new-line character
`fgetl`	reads a line from file including new-line character
`frewind`	rewinds a file
`fseek`	sets the file position indicator
`ftell`	gets the current file position indicator
`ferror`	inquires file I/O error status.

You are likely to use only the first six commands in the list for file I/O. For most purposes `fopen`, `fprintf`, and `fclose` should suffice. For a complete description of these commands see the on-line help or consult the Reference Guide [2] or a C-language reference book[7].

Here is a simple example that uses `fopen`, `fprintf`, and `fclose` to create and write formatted data to a file:

```
% TEMTABLE - generates and writes a temperature table
% Script file to generate a Fahrenheit-Celsius
% temperature table. The table is written in
% a file named 'Temperature.table'.
% -----------------------------------------------
F=-40:5:100;            % take F=[-40 -35 -30 .. 100]
C=(F-32)*5/9;           % compute corresponding C
t=[F;C];                % create a matrix t (2 rows)
fid = fopen('Temperature.table','w');
fprintf(fid,'  Temperature Table\n ');
fprintf(fid,' ~~~~~~~~~~~~~~~~ \n');
fprintf(fid,'Fahrenheit    Celsius \n');
fprintf(fid,'  %4i       %8.2f\n',t);
fclose(fid);
```

In the above script file, the first I/O command, `fopen`, opens a file `Temperature.table` in the *write* mode (specified by `'w'` in the command) and assigns the *file identifier* to `fid`. The following `fprintf` commands use `fid` to write the strings and data to that file. The data is formatted according to the specifications in the string argument of `fprintf`. In the command above, `\n` stands for *new line*, `%4i` stands for an *integer* field of width 4, and `%8.2f` stands for a *fixed point* field of width 8 with 2 digits after the decimal point.

The output file, `Temperature.table`, is shown below. Note that the data matrix `t` has two rows whereas the output file writes the matrix in two columns. This is because `t` is read columnwise and then written in the format specified (two values in a row).

```
        Temperature Table
        ~~~~~~~~~~~~~~~~~

   Fahrenheit      Celsius
      -40          -40.00
      -35          -37.22
      -30          -34.44
      -25          -31.67
      -20          -28.89
      -15          -26.11
      -10          -23.33
       -5          -20.56
        0          -17.78
        :             :
        :             :
       95           35.00
      100           37.78
```

4.4 Advanced Data Objects

In MATLAB 5 some new important data objects have been introduced, namely *structures* and *cells*. Also, the most familiar data object, a matrix, has been transformed — now it can be *multidimensional*. Although a detailed discussion of these objects and their applications is beyond the scope of this book, in this section, we give enough introduction and examples to get you started. At first, these objects look very different from the MATLAB's heart and soul, a matrix. The fact, however, is that these objects, as well as a matrix, are just special cases of the fundamental data type, the *array*. As you become more comfortable with these objects, they fall into their right places and you start working with them just as you would with vectors and matrices.

4.4.1 Multidimensional matrices

MATLAB 5 supports multidimensional matrices. You can create n dimensional matrices by specifying n indices. The usual matrix creation functions, zeros, ones, rand, and randn, accept n indices to create such matrices. For example,

A = zeros(4,4,3)   initializes a $4 \times 4 \times 3$ matrix A
B = rand(2,4,5,6)   creates a $2 \times 4 \times 5 \times 6$ random matrix B

So, how do we think about the other (the 3rd, 4th, etc.) dimensions of the matrix? You can write a 2-D matrix on a page of a notebook. Think of the third dimension as the different pages of the notebook. Then matrix A (=zeros(4,4,3)) occupies three pages, each page having a 4×4 matrix of 0's. Now, if we want to change the entry in the 4th row and 4th column of the matrix on page 2 to a number 6, we type A(4,4,2) = 6. Thus the usual rules of indexing apply; we only have to think which matrix in the stack we are accessing. Now, once we have moved out of the page (a 2-D matrix), we have no limitation — we can have several notebooks (4th dimension), several bookcases full of such notebooks (5th dimension), several rooms full of such bookcases (6th dimension), and so forth and so on. The multidimensional indexing makes it easy to access elements in any direction you wish — you can get the first element of the matrices on each page, from each notebook, in each bookcase, from all rooms, with C(1,1,:,:,:,:). The usual indexing rules apply in each dimension, therefore, you can access submatrices with index ranges just as you would for a 2-D matrix.

When you operate on multidimensional matrices, you have to be careful. All linear algebra functions will work only on 2-D matrices. You cannot multiply two matrices if they are 3-D or higher dimensional matrices. Matrix multiplication is not defined for such matrices. All array operations (element by element operation) are, however, valid for any dimensional matrix. Thus, 5*A, sin(A), or log(A) are all valid commands for matrix A of any dimension. Similarly, if A and B are two matrices of the same dimension then A+B or A-B is valid irrespective of the dimensionality of A and B.

4.4.2   Structures

A *structure* is a data construct with several named *fields*. Different fields can have different types of data, but a single field must contain data of the same type.

A structure is like a record. One record (structure) can contain information (data) about various things under different heads (fields). For example, you could maintain a record book with one page devoted to each of your relatives. You could list information about them under the heads *relationship, address, name of children, date of birth of all children*, etc. Although, the headings are the same in each record, the information contained under a heading could vary in length from record to record. To implement this record keeping in MATLAB, what you have is a structure. But, what's more is that you are not limited to a serial array of record pages (your record book), you can have a multidimensional structure.

Let us now look at an example of a structure. We are going to make a structure called `FallSem` with fields `course`, `prof`, and `score`. We wish to record the names of courses, names of corresponding professors, and your performance in tests in those courses. Here is one way to do this:

```
FallSem.course = 'cs101';
FallSem.prof = 'Turing';
FallSem.score = [80 75 95];
```

Thus, *fields* are indicated by adding their names after the name of the structure, with a dot separating the two names. The fields are assigned values just as any other variable in MATLAB. In the above example structure, `FallSem`, the fields `course` and `prof` contain character strings, while `score` contains a vector of numbers. Now, how do we generate the record for the next course? Well, structures as well as their fields can be multidimensional. Therefore, we can generate the next record in a few different ways:

Multiple records in a *Structure Array*: We can make the structure `FallSem` to be an array (in our example a vector will suffice), and then store a complete record as one element of the array:

```
FallSem(2).course = 'phy200';      FallSem(3).course = 'math211';
FallSem(2).prof = 'Fiegenbaum';    FallSem(3).prof = 'Ramanujan';
FallSem(2).score = [72 75 78];     FallSem(3).score = [85 35 66];
```

Thus, we have created a structure array `FallSem` of size 1×3. Each element of `FallSem` can be accessed just as you access an element of an usual array — `FallSem(2)` or `FallSem(1)`, etc. By typing `FallSem(1)`, you get the values of all the fields, along with the field names. You can also access individual fields, for example, `FallSem(1).prof` or `FallSem(1).score(3)`.

See Fig. 4.1 for example output.

Note: In a structure array, each element must have the same number of fields. Each field, however, can contain data of different sizes. Thus, `FallSem(1).score` can be a 3 element long row vector while `FallSem(2).score` can be a 5 element long column vector.

Multiple records in *Field Arrays*: For the example we have chosen, we could store multiple records in a single structure (i.e., keep `FallSem` as a 1×1 structure) by making the fields to be arrays of appropriate sizes to accommodate the records:

```
FallSem.course = char('cs101','phy200','math211');
FallSem.prof = char('Turing','Fiegenbaum','Ramanujan');
FallSem.score = [80 75 95; 72 75 78; 85 35 66];
```

```
>> FallSem.course = 'cs101';
>> FallSem.prof = 'Turing';
   FallSem.score = [80 75 95];
```
Create a structure FallSem with three fields -- course, prof, and score.

```
>> FallSem

   FallSem =
       course: 'cs101'
         prof: 'Turing'
        score: [80 75 95]
```
When queried, MATLAB shows the entire structure.

```
>> FallSem(2).course = 'phy200';   FallSem(3).course = 'math211';
>> FallSem(2).prof = 'Fiegenbaum'; FallSem(3).prof = 'Ramanujan';
>> FallSem(2).score = [72 75 78];  FallSem(3).score = [85 35 66];
```

```
>> FallSem

   FallSem =
   1x3 struct array with fields:
       course
       prof
       score
```
After adding two more records FallSem becomes an array, and when queried, MATLAB now gives structural info about the structure.

```
>> FallSem(2).course

   ans =
   phy200
```
Use array index on the structure to access its elements.

```
>> FallSem(3).score(1)

   ans =
       85
```
You can use index notation for the structure as well as its fields.

```
>> FallSem.score

   ans =
       80      75      95
   ans =
       72      75      78
   ans =
       85      35      66
```
When no index is specified for the structure, MATLAB displays the field values of all records.

```
>> for k=1:3,
       all_scores(k,:) = FallSem(k).score;
   end
```
To assign values from a field of several records use a loop.

Figure 4.1: A tutorial lesson on structures.

In this example, the function char is used to create separate rows of character strings from the input variables. Here, FallSem is a 1×1 structure, but the field course is a 3×7 character array, prof is a 3×10 character array and score is a 3×3 array of numbers.

Well, this example works out nicely because we could create a column vector for course names, a column vector for professors' names, and a matrix for scores where each row corresponds to a particular course. What if we had the third record as a matrix for each course? We could still store the record in both ways mentioned above. While the first method of creating a structure array seems to be the easiest way, we could also use the second method and have the third field score to be a 3-dimensional matrix.

Creating structures

In the examples above, we have already seen how to create structures by direct assignment. Just as you create a matrix by typing its name and assigning values to it — A = [1 2; 3 4]; — you can create a structure by typing its name along with a field and assigning values to the field. This is what we did in the examples. The other way of creating a structure is with the struct function. The general syntax of struct is

```
str_name = struct('fieldname1', 'value1', 'fieldname2', 'value2', ...)
```

Thus, the structure FallSem created above could also be created using the struct function as follows:

As a single structure:

```
FallSem = struct('course',char('cs101','phy200','math211'),...
            'prof', char('Turing','Fiegenbaum','Ramanujan'),...
            'score',[80 75 95; 72 75 78; 85 35 66]);
```

As a structure array:

```
Fall_Sem = [struct('course','cs101','prof','Turing',...
                'score',[80 75 95]);
            struct('course','phy200','prof','Fiegenbaum',...
                'score',[72 75 78]);
            struct('course','math211','prof','Ramanujan',...
                'score',[85 35 66])];
```

Note that this construction creates a 3×1 (a column vector) structure array Fall_Sem.

Manipulating structures

Manipulation of structures is similar to the manipulation of general arrays
— access elements of the structure by proper indexing and manipulate their
values. There is, however, a major difference; you cannot assign all values
of a field *across* a structure array to a variable with colon range specifier.
Thus, if `Fall_Sem` is a 3×1 structure array, then

`Fall_Sem(1).score(2)`	is valid and gives the second element of `score` from the first record of `Fall_Sem`
`Fall_Sem(1).score(:)`	gives all elements of `score` from `Fall_Sem(1)`, same as `Fall_Sem(1).score`
`a = Fall_Sem(:).score`	is invalid, does not *assign* `score` from all records to `a`, though the command `Fall_Sem(:).score` or `(Fall_Sem.score)` displays scores from all records.

So, although you can *see* (on the screen) field values across multiple
records with `Fall_Sem(:).score`, you have to use a loop construction to
assign the values to a variable:

```
for k=1:3, all_scores(k,:)  = Fall_Sem(k).score; end
```

The assignment cannot be done directly with the colon operator because
the values of fields from several records are treated as separate entities. The
field contents are also allowed to be of different sizes. Therefore, although
you can use a *for*-loop for assignment, you have to take extra care to ensure
that the assignment makes sense. For example, in the `for`-loop above, if
`Fall_Sem(2).score` has only two test scores then the assignment above will
produce an error.

It is clear from the above examples that you can use indices for structures as well as fields for accessing information. Here, we have used only
character arrays and number arrays in the fields. You can, however, also
have structures inside structures. But, the level of indexing becomes quite
involved and requires extra care if you have nested structures.

There are also several functions provided to aid manipulation of structures — `fieldnames`, `setfield`, `getfield`, `rmfield`, `isfield`, etc. The
names of most of these functions are suggestive of what they do. To get the
correct syntax, please see the on-line help on these functions.

4.4.3 Cells

A cell is the most versatile data object in MATLAB. It can contain any type
of data — an array of numbers, strings, structures, or cells. An array of cells
is called a *cell array*.

You can think of a cell array as an array of data containers. Imagine putting nine empty boxes on the floor in three rows and three columns. Let us call this arrangement a 3×3 cell, **my_riches**. Now, let us put clothes in one box, shoes in the second box, books in the third box, a computer in the fourth box, all greenbacks and coins in the fifth box, and so on. Each box is a container just like any other box, however, the contents of each are different. Thus, we have an arrangement that can accommodate all kinds of things in the containers and still look superficially similar. Now replace the boxes with MATLAB data containers, **cells**, and replace the shoes with a *matrix*, the clothes with a **string array**, the books with a **structure**, the computer with another **cell** (many little boxes inside the big box) and so on, and what you have got, at the topmost level, is a MATLAB data object called *a cell*.

Let us first create a cell, put some data in it, and then discuss various aspects of cells using this cell as an example:

```
C = cell(2,2);            % create a 2 by 2 cell
                          % cell(2) will do the same
C{1,1} = rand(3);         % put a 3x3 random matrix in the 1st box
C{1,2} = char('john','raj'); % put a string array in the 2nd box
C{2,1} = Fall_Sem;        % put a structure in the 3rd box
C{2,2} = cell(3,3);       % put a 3x3 cell in the 4th box
```

In this example, creating a cell superficially looks like creating an ordinary array, but there are some glaring differences. Firstly, the contents are as varied as we want them to be. Secondly, there are curly braces, instead of parenthesis, on the left side of the assignment. So, what are those curly braces for? Could we use parenthesis instead?

A cell is different from an ordinary (number) array in that it distinguishes between the (data) containers and the contents, and it allows access to both, separately. When you treat a cell as an array of containers, without paying any attention to the contents, the cell behaves just as an array and you can access a container with the familiar indexing syntax, **C(i,j)**. What you get is the container located at the *i*th row and *j*th column. The container will carry a label that will tell you whether the contents are **double**, **char**, **struct**, or **cell**, and of what dimension. If you want to access the contents of a container, you have to use the special *cell-content-indexing* — indices enclosed within curly braces. Thus, to see the 3×3 random matrix in **C(1,1)**, you type **C{1,1}**. See Fig. 4.2 for some examples.

Creating cells

We have already discussed above how to create cells using the **cell** function. We can also create cells directly:

```
C = {rand(3)   char('john','raj');   Fall_Sem   cell(3,3)};
```

```
>> Fall_Sem =
    [struct('course','cs101','prof','Turing','score',[80 75 95]);
     struct('course','phy200','prof','Fiegenbaum','score',[72 75 78]);
     struct('course','math211','prof','Ramanujan','score',[85 35 66 )];
```

Create a 3 by 1 structure `Fall_Sem` with 3 fields

```
>> C = cell(2,2);
>> C{1,1} = rand(3);
>> C{1,2} = char('john','raj');
>> C{2,1} = Fall_Sem;
>> C{2,2} = cell(3,3);
```

Create a 2 by 2 cell C and put different types of data in the individual containers of C.

```
>> C

   C =

        [3x3 double]    [2x4 char]

        [3x1 struct]    {3x3 cell}
```

When queried, MATLAB shows what type of contents do the containers of C have.

```
>> C{1,2}

   ans =
   john
   raj
```

Use *content-indexing* , {i,j}, on the cell to access the contents of the container at location (i,j).

```
>> C(1,2)

   ans =
       [2x4 char]
```

Use array index notation to access a container (but not its contents)

```
>> C{1,2}(1,:)

   ans =
   john
```

You can use multiple index notation to access a particular data in a particular container.

```
>> C{2,1}(3).prof

   ans =
   Ramanujan
```

You can also combine the structure notation with cell indices.

```
>> C{2,1}(3).score(3)

   ans =
       66
```

This command locates `Fall_Sem(3).score(3)` in C.

Figure 4.2: A tutorial lesson on cells.

This example illustrates that the curly braces are to a cell what square brackets are to an ordinary array when used on the right side of an assignment statement.

Manipulating cells

Once you get used to curly braces versus parenthesis on both sides of an assignment statement, manipulating cells is just as easy as manipulating an ordinary array. However, there is much more fun to be had with cell indexing, the type of indexing-fun you have not had before. Create the cell C given in the example above along with the structure Fall_Sem given on page 109. Now try the following commands and see what they do.

```
C(1,1) = {[1 2; 3 4]};
C{1,2}(1,:)
C{2,1}(1).score(3) = 100;
C{2,1}(2).prof
C{2,2}{1,1} = eye(2);
C{2,2}{1,1}(1,1) = 5;
```

There are several functions available for cell manipulation. Some of these functions are `cellstr`, `iscellstr`, `cell2struct`, `struct2cell`, `iscell`, `num2cell`, etc. See the on-line help on these functions for correct syntax and usage.

Two cell functions deserve special mention. These functions are

`celldisp`	displays the contents of a cell on the screen
	works recursively on cells inside cells
`cellplot`	plots the cell array schematically

The result of `cellplot(C)` is shown in Fig. 4.3 for cell C created in Fig. 4.2. The contents of each cell are shown schematically as arrays of appropriate sizes. The non-empty array elements are shown shaded.

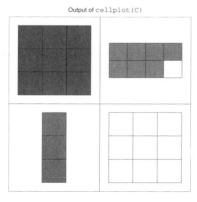

Figure 4.3: Cells can be displayed schematically with the function `cellplot`.

EXERCISES

1. **A script file to compute sine series:** Write a script file named `sineseries.m` that computes the value of $\sin(x)$ at a given x using n terms of the series expansion of sine function:

$$\sin(x) = x - \frac{x^3}{3!} + \frac{x^5}{5!} - \cdots = \sum_{k=1}^{n} (-1)^{k-1} \frac{x^{2k-1}}{(2k-1)!}$$

   Follow the steps given below.

   - First, query MATLAB to see if the name 'sineseries' is already taken by some variable or function with the command `exist(sineseries)`. What does the MATLAB response mean? [Hint, see on-line help on `exist`.]
   - Include the following line as the header (H1 line) of your script file.
     `%SINESERIES: computes sin(x) from series expansion`
     Now code the formula so that it computes the sum of the series for a given scalar x and a given integer n.
   - Save the file. Type `help sineseries` to see if MATLAB can access your script file. Now, compute $\sin(\pi/6)$ with $n = 1, 5, 10$, and 20. Compare the results. Do the same for some other x of your choice.

2. **A function file to compute sine series:** Take the script file written above and convert it into a function file using the following steps.

   - Name the function `sine_series` and modify the H1 line appropriately.
   - Let x and n be the input to your function and y (the sum) be the output.
   - Save the function and execute it to see that it works and gives the same output as the script file above.
   - Modify the function to include more on-line help on how to run the function.
   - Modify the function so that it can accept a vector x and give an appropriate y.
   - Modify the function to include a check on the input n. The function should proceed only if $n > 0$ is an integer, otherwise it should display an error message.
   - Provide for an optional output `err` which gives the % error in y when compared to `sin(x)`. [Hint: use conditional statement on `nargout` for optional output.]
   - Modify the function so that it takes a default value of $n = 10$ if the user does not specify n. [Hint: use `nargin`.]
   - Execute the function to check all features you have added.

3. **A function as an input to another function:** There are several ways in which a function can be passed in the input argument list of another function. The function to be used in the input list can be written as an `inline` function (see Section 3.3 on page 67) or it can be coded in a function file. How the function is passed in the input list depends on how it is coded.

Code the function $y(x) = \frac{\sin(x)}{x}$ as an inline function `sinc` and in a function file called `sincfun.m`. You will this function in the input list of `ezplot` (see page 78 for `ezplot`) in various ways in the following instructions.

- Use the inline function `sinc` in the input list of `ezplot` to plot the function over the default domain.
- Use the function `sincfun` as a string in the input list of `ezplot` to plot the function over the default domain.
- Create a function handle for `sincfun` and use the handle in the input list of `ezplot` to plot the function over the default domain.

4. **Write subfunctions:** In this exercise you will write a simple function to find $\sin(x)$ and/or $\cos(x)$ from series expansion using two subfunctions—`sine_series` and `cosine_series`.

- Write a function named `cosine_series` to evaluate the cosine series (follow Exercise 2 above)

$$\cos(x) = 1 - \frac{x^2}{2!} + \frac{x^4}{4!} - \cdots = \sum_{k=1}^{n} (-1)^{k-1} \frac{x^{2(k-1)}}{2(k-1)!}$$

Execute and test the function to make sure it works correctly.
- Write a new function named `trigseries` as follows.
  - The input list of the function should include x, n, and another string variable '*options*'. The user can specify '`sin`', '`cos`', or '`both`' for '*options*'. The default should be '`both`'. The output list should include y and *err* as discussed in Exercise 2.
  - The function `trigseries` should call `sine_series` and `cosine_series` as appropriate to compute y and *err*, depending on what the user specifies in '*options*'. Implement this call sequence with `switch` using `options` as the switch. Program the output
  - y and *err* to be two column arrays if the user asks for '`both`' in '*options*' or as the default.
- Copy and paste the functions `sine_series` and `cosine_series` as subfunctions below the function `trigseries` in the same file.
- Delete the original files `sine_series.m` and `cosine_series.m` or re-name them something else. Test `trigseries` with various input and '*options*'. Make sure it works correctly. Now execute `trigseries` with only one input `x = [0 pi/6 pi/4 pi/3 pi/2]`. Do you get reasonable answers?

5. **Profile a function:** Profile the function `trigseries`, developed above in Exercise 4, taking a vector x of 100 equally spaced points between 0 and π as the only input.

6. **Recursion:** Write a function to compute $n!$ using recursion. (Note that this is not the most efficient way to compute $n!$. However, it is conceptually a recursive calculation that is easy to implement and test recursion in MATLAB.)

5. Applications

5.1 Linear Algebra

5.1.1 Solving a linear system

Solving a set of linear algebraic equations is easy in MATLAB. It is, perhaps, also the most used computation in science and engineering. Solving a system of equations on a computer is now as basic a task as doing scientific calculations on a calculator. Let us see how MATLAB makes this task easy and pleasant.

We will solve a set of linear algebraic equations given below. To solve these equations, no prior knowledge of matrix methods is required. The first two steps outlined below are really basic for most people who know a little bit of linear algebra. We will consider the following set of equations for our example.

$$5x = 3y - 2z + 10$$
$$8y + 4z = 3x + 20$$
$$2x + 4y - 9z = 9$$

Step-1: Rearrange equations: Write each equation with all unknown quantities on the left-hand side and all known quantities on the right hand side. Thus, for the equations given above, rearrange them such that all terms involving x, y, and z are on the left side of the equal sign:

$$
\begin{aligned}
5x - 3y + 2z &= 10 \\
-3x + 8y + 4z &= 20 \\
2x + 4y - 9z &= 9
\end{aligned}
\tag{5.1}
$$

Step-2: Write the equations in matrix form: To write the equation in the matrix form $[\mathbf{A}]\{\mathbf{x}\} = \{\mathbf{b}\}$ where $\{\mathbf{x}\}$ is the vector of unknowns, you have to arrange the unknowns in vector $\mathbf{x}$, the coefficients of the unknowns in matrix $\mathbf{A}$ and the constants on the right hand side of the equations in vector $\mathbf{b}$. In this particular example, the unknown vector is

$$\mathbf{x} = \begin{bmatrix} x \\ y \\ z \end{bmatrix},$$

the coefficient matrix is

$$\mathbf{A} = \begin{bmatrix} 5 & -3 & 2 \\ -3 & 8 & 4 \\ 2 & 4 & -9 \end{bmatrix},$$

and the known constant vector is

$$\mathbf{b} = \begin{bmatrix} 10 \\ 20 \\ 9 \end{bmatrix}.$$

Note that the columns of $\mathbf{A}$ are simply the coefficients of each unknown from all the three equations. Now you are ready to solve this system in MATLAB.

Step-3: Solve the matrix equation in MATLAB: Enter the matrix $\mathbf{A}$ and vector $\mathbf{b}$, and solve for vector $\mathbf{x}$ with x = A\b (note that the \ is different from the division /):

For on-line help type:
`help slash`

```
>> A = [5 -3 2; -3 8 4; 2 4 -9];   % Enter matrix A
>> b = [10; 20; 9];                % Enter column vector b
>> x = A\b                         % Solve for x
x =
    3.4442
    3.1982
    1.1868

>> c = A*x                         % check the solution
c =
   10.0000
   20.0000
    9.0000
```

The backslash (\) or the left division is used to solve a linear system of equations $[A]\{x\} = \{b\}$. For more information type: `help slash`.

5.1.2   Gaussian elimination

In introductory linear algebra courses, we learn to solve a system of linear algebraic equations by *Gaussian elimination*. This technique requires forming a rectangular matrix that contains both the coefficient matrix $\mathbf{A}$ and the known vector **b** in an *augmented matrix*. Gauss-Jordan reduction procedure is then used to transform the augmented matrix to the so called *row reduced echelon form*. MATLAB has a built-in function, `rref` that does precisely this reduction, i.e., transforms the matrix to its row reduced echelon form. For example, consider eqn. (5.1) with the coefficient matrix **A** and the known vector **b** formed as above. To solve the equations with `rref`, type the following commands:

```
C = [A  b];      % form the augmented matrix
Cr = rref(C);    % row reduce the augmented matrix
```

The last column of `Cr` is the solution `x`. You may like to use `rref` for checking homework solutions; it is not very useful for anything else.

5.1.3 Finding eigenvalues & eigenvectors

The omnipresent eigenvalue problem in scientific computation shows up as

$$\mathbf{A}\,\mathbf{v} = \lambda\mathbf{v} \tag{5.2}$$

λ is a scalar. The problem is to find λ and **v** for a given **A** so that eqn. (5.2) is satisfied. By hand, the solution is usually obtained by first solving for the n eigenvalues from the determinant equation $|\mathbf{A} - \lambda\mathbf{I}| = 0$, and then solving for the n eigenvectors by substituting the corresponding eigenvalues in eqn. (5.2), one at a time. On pencil and paper, the computation requires a few pages even for a 3×3 matrix,[1] but it is a 1.2 inch long command in MATLAB (excluding entering the matrix)! Here is an example:

Step-1: Enter matrix A and type `[V,D] = eig(A)`. (See Fig. 5.1 on page 120).

Step-2: Extract what you need: In the output-list, `V` is an $n \times n$ matrix whose columns are eigenvectors and `D` is an $n \times n$ diagonal matrix which has the eigenvalues of `A` on its diagonal. The function `eig` can also be used with one output argument, e.g., `lams = eig(A)`, in which case the function gives only the eigenvalues in the vector `lams`.

After computing the eigenvalues and eigenvectors, you could check a few things if you wish: Are the eigenvalues and eigenvectors ordered in the output? Check by substituting in eqn. (5.2). For example, let us check the second eigenvalue and second eigenvector and see if, indeed, they satisfy $Av = \lambda v$.

```
>> v2 = V(:,2);     % extract the second column from V
>> lam2 = D(2,2);   % extract the second eigenvalue from D
>> A*v2 - lam2*v2   % check the difference between A*v and lambda*v
```

Here, we have used two commands for clarity. You could, of course, combine them into one: `A*V(:,2)-D(2,2)*V(:,2)`.

---

[1]If A is bigger than 4×4, you have to be insane to try to solve it by hand, for a 4×4 matrix, you are either borderline insane or you live in a civilization without computers.

```
>> A = [5 -3  2; -3  8  4; 4  2  -9];
>> [V,D] = eig(A)
```

V =

Here V is a matrix containing the eigenvectors of A as its columns. For example, the first column of V is the first eigenvector of A.

```
        0.1725        0.8706       -0.5375
        0.2382        0.3774        0.8429
       -0.9558        0.3156       -0.0247
```

D =

D is a matrix that contains the eigenvalues of A on its diagonal.

```
      -10.2206             0             0
             0        4.4246             0
             0             0        9.7960
```

Figure 5.1: Finding eigenvalues and eigenvectors of a matrix.

5.1.4   Matrix factorizations

MATLAB provides built-in functions for several matrix factorizations (decompositions):

1. **LU factorization:** The name of the built-in function is lu. To get the LU factorization of a square matrix A, type the command

$$[L,U] = lu(A);$$

MATLAB returns a lower triangular matrix L and an upper triangular matrix U such that L*U = A. It is also possible to get the permutation matrix as an output (see the on-line help on lu).

2. **QR factorization:** The name of the built-in function is qr. Typing the command

$$[Q,R] = qr(A);$$

returns an orthogonal matrix Q and an upper triangular matrix R such that Q*R = A. For more information, see the on-line help.

3. **Cholesky factorization:** If you have a positive definite matrix A you can factorize the matrix with the built-in function chol. The command

$$R = chol(A);$$

produces an upper triangular matrix R such that R'*R = A for a positive definite A.

4. **Singular Value Decomposition (svd):** The name of the built-in function is svd. If you type

$$[U,D,V] = svd(A);$$

MATLAB returns two orthogonal matrices U and V, and a diagonal matrix D, with the *singular values* of A as the diagonal entries, such that U*D*V = A.

If you know which factorization to use, then all you need to know here is the syntax of the corresponding function. If you do not know where and how these factorizations are used, then this is not the right place to learn it; look into your favorite books on linear algebra.[2]

5.1.5 Advanced topics

MATLAB's main strength is its awesome suite of linear algebra functions. There are hundreds of functions that aid in solution of linear algebra problems, from very basic problems to very advanced ones. For example, there are more than ten eigenvalue related functions. Type lookfor eigenvalue to see a list of these functions. Among these functions, eigs deserves a special mention. This function has been added in MATLAB 5.2 to help in finding just a few eigenvalues and eigenvectors, a requirement that frequently arises in large scale problems.

For on-line help type:
help sparfun

There is a separate suite of functions for sparse matrices and computations with these matrices. These functions include:

Category	Example functions
Elementary matrix functions	speye, sprand, spdiags
Full to sparse conversion	sparse, full, spconvert
Utility functions	nnz, nzmax, spalloc, spy
Reordering algorithm	colmmd, symmd, symrcm
Linear algebra	eigs, svds, luinc, cholinc
Linear equations	pcg, bicg, cgs, qmr

Please see the on-line help with help sparfun for list of these functions and their brief descriptions. Of course, detailed on-line help is available on each function as well.

MATLAB 5 also provides several functions for operations on graphs. These functions are listed under the **sparfun** category.

---

[2]Some of my favorites: *Linear Algebra and Its Applications* - Strang, Saunders HBJ College Publishers; *Matrix Computations* - Golub & Van loan, The Johns Hopkins University Press; *Matrix Analysis* - Horn & Johnson, Cambridge University Press.

5.2   Curve Fitting and Interpolation

5.2.1   Polynomial curve fitting on a fly

Curve fitting is a technique of finding an algebraic relationship that "best" (in a *least squares* sense) fits a given set of data. Unfortunately, there is no magical function (in MATLAB or otherwise) that can give you this relationship if you simply supply the data. You have to have an idea of what kind of relationship might exist between the input data (x_i) and the output data (y_i). However, if you do not have a firm idea but you have data that you trust, MATLAB can help you in exploring the best possible fit. MATLAB 6 includes *Basic Fitting* in its **Figure** window's **Tools** menu that lets you fit a polynomial curve (up to 10th order) to your data on a fly. It also gives you options of displaying the residual at the data points and computing the norm of the residuals. This can help in comparing different fits and then selecting the one that makes you happy. Let us take an example and go through the steps.

Example-1: Straight-line (linear) fit:

Let us say that we have the following data for x and y and we want to get the best linear (straight-line) fit through this data.

x	5	10	20	50	100
y	15	33	53	140	301

Here is all it takes to get the best linear fit, along with the equation of the fitted line.

Step-1: Plot raw data: Enter the data and plot it as a scatter plot using some marker, say, circles.

```
x = [5 10 15 20 50 100];        % x-data
y = [15 33 53.39 140 301];      % y-data
plot(x,y,'o')                   % plot x vs y using circles
```

Step-2: Use built-in *Basic Fitting* to do a linear fit: Go to your figure window, click on **Tools**, and select **Basic Fitting** from the pull down menu (see Fig. 5.2). A separate window appears with *Basic Fitting* options.

Step-3: Fit a linear curve and display the equation: Check the boxes for **linear** and **Show equations** from the *Basic Fitting* window options. The best fitted line as well as its equation appear in the figure window. That is it, you are done. The result is shown in Fig. 5.3.

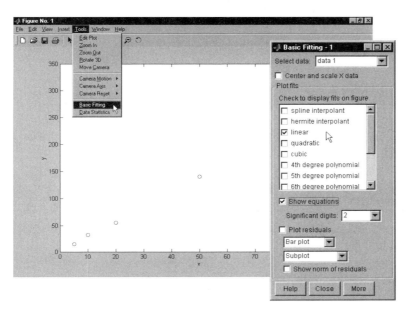

Figure 5.2: First plot the raw data. Then select **Basic Fitting** from the **Tools** menu of the figure window. A separate **Basic Fitting** window appears with several options for polynomial curve fits. Check appropriate boxes in this window to get the desired curve fit and other displays, such as the fitted equation.

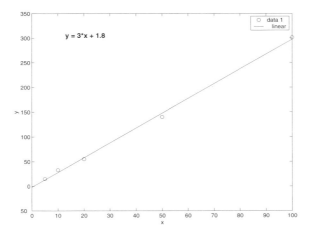

Figure 5.3: Linear curve fit through the data using **Basic Fitting** from **Tools** menu of the figure window.

Example-2: Comparing different fits:

Let us take another example where we try two different fits, quadratic and cubic, for the same data and do a comparison to figure out which one is better. We first create some x and y data.

```
x = 0 : pi/30 : pi/3;            % x-data
y = sin(x) + rand(size(x))/100  % y-data (corrupted sine)
```

Step-1: Plot raw data: We already have the x and y data. So, go ahead and plot the raw data with `plot(x,y,'o')`.

Step-2: Use *Basic Fitting* to do a quadratic and a cubic fit: Go to your figure window, click on **Tools**, and select **Basic Fitting** from the pull down menu (as in Example-1 above). In the *Basic Fitting* window, check **quadratic** and **cubic** boxes. In addition, check the boxes for **Show equations**, **Plot residuals**, and **Show norm of residuals**.

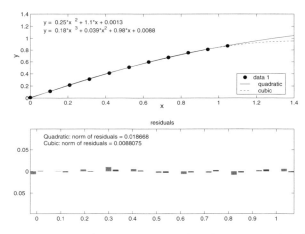

Figure 5.4: A comparison of quadratic and cubic curve fits for some data using **Basic Fitting** from **Tools** menu of the figure window.

The result is shown in Fig. 5.4. The residual at each data point is plotted in the lower subplot for each fit and the norm of the residual is also shown on the plot. Note that although the two curves are visually almost indistinguishable over the range of the data, the norm of the residuals for the cubic fit is an order of magnitude lower than that of the quadratic fit, and therefore, you may like to chose the cubic fit in this case.

For on-line help type:
`help polyfun`

5.2.2 Do it yourself: curve fitting using polynomial functions

It may be worth understanding how these curve fits work. In MATLAB, it is fairly easy to do polynomial curve fits using built-in polynomial functions and get the desired coefficients. If you would like to understand it, please read on, otherwise you may like to skip this section.

The simplest relationship between two variables, say x and y, is a linear relationship: $y = mx + c$. For a given set of data points (x_i, y_i), the problem is to find m and c such that $y_i = mx_i + c$ best fits the data. For data points which are not linearly related, you may seek a polynomial relationship

$$y_i = a_k x_i^k + a_{k-1} x_i^{k-1} + \cdots + a_2 x_i^2 + a_1 x_i + a_0$$

or an exponential relationship $y_i = c_1 e^{c_2 x_i}$ or even more complicated relationships involving logarithms, exponentials, and trigonometric functions.

For polynomial curve fits, of any order n, the problem is to find the $(n + 1)$ coefficients $a_n, a_{n-1}, a_{n-2}, \cdots a_1$, and a_0, from the given data of length $(n + 1)$ or more. MATLAB provides an easy way — through the built-in functions `polyfit` and `polyval` .

`polyfit:` Given two vectors x and y, the command a = `polyfit(x,y,n)` fits a polynomial of order n through the data points (x_i, y_i) and returns $(n + 1)$ coefficients of the powers of x in the row vector a. The coefficients are arranged in the decreasing order of the powers of x, i.e., a = $[a_n \quad a_{n-1} \quad \cdots \quad a_1 \quad a_0]$.

`polyval:` Given a data vector x and the coefficients of a polynomial in a row vector a, the command y = `polyval(a,x)` evaluates the polynomial at the data points x_i and generates the values y_i such that

$$y_i = a(1)x_i^n + a(2)x_i^{n-1} + \cdots + a(n)x + a(n+1).$$

Here the length of the vector a is $n + 1$ and, consequently, the order of the evaluated polynomial is n. Thus if a is 5 elements long vector, the polynomial to be evaluated is automatically ascertained to be of 4th order.

Both `polyfit` and `polyval` use an optional argument if you need error estimates. To use the optional argument, see the on-line help on these functions.

Example-1: Straight-line (linear) fit:

The following data is obtained from an experiment aimed at measuring the spring constant of a given spring. Different masses m are hung from the spring and the corresponding deflections δ of the spring from its unstretched configuration are measured. From physics, we know that $F = k\delta$ and here $F = mg$. Thus, we can find k from the relationship $k = mg/\delta$. Here, however, we are going to find k by plotting the experimental data, fitting the best straight-line (we know that the relationship between δ and F is linear) through the data, and then measuring the slope of the best-fit line.

m(g)	5.00	10.00	20.00	50.00	100.00
δ(mm)	15.5	33.07	53.39	140.24	301.03

Fitting a straight line through the data means we want to find the polynomial coefficients a_1 and a_0 (a first order polynomial) such that $a_1 x_i + a_0$ gives the "best" estimate of y_i. In steps, we need to do the following.

Step-1: Find the coefficients a_k's:

```
a = polyfit(x,y,1)
```

Step-2: Evaluate y at finer (more closely spaced) x_j's using the fitted polynomial:

```
y_fitted = polyval(a,x_fine)
```

Step-3: Plot and see. Plot the given data as points and fitted data as a line.

```
plot(x,y,'o',x_fine,y_fitted);
```

As an example, the following script file shows all the steps involved in making a straight line fit through the data for the spring experiment given above and finding the spring constant. The resulting plot is shown in Fig. 5.5

```
m=[5 10 20 50 100];               % mass data (g)
d=[15.5 33.0755.39 140.24 301.03]; % displacement data (mm) g=9.81;
F=m/1000*g;                       % compute spring force (N)
a=polyfit(d,F,1);                 % fit a line (1st order polynomial)
d_fit=0:10:300;                   % make a finer grid on d
F_fit=polyval(a,d_fit);           % evaluate the polynomial at new points
plot(d,F,'o',d_fit,F_fit)         % plot data and the fitted curve
xlabel('Displacement (mm)'),ylabel('Force (N)') k = a(1);
                                  % Find the spring constant
text(100,.32,['\leftarrow Spring Constant K = ',num2str(k)]);
```

Comments: While you can do a polynomial curve fit of any order, you should have a good reason, possibly based on the physical phenomena behind the data, for trying a particular order. Blindly fitting a higher order polynomial is often misleading. The most common and trustworthy curve fit, by far, is the straight line. If you expect the data to have exponential relationship, convert the data to a log scale and then do a linear curve fit. The result is a lot more trustworthy than an arbitrary 6th or 10th order polynomial fit. As a rule of thumb, polynomial curve fits of order higher than 4 or 5 are rarely required.

5.2.3   Least squares curve fitting

The technique of least squares curve fit can easily be implemented in MATLAB, since the technique results in a set of linear equations that need to be solved.

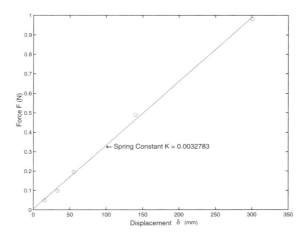

Figure 5.5: A straight line fit (a polynomial fit of order one) through the spring data.

Here, we will not discuss the details of the technique itself because of limitation of space as well as intent of the book. You may like to consult a book on numerical methods if you do not know the underlying principles.

Most of the curve fits we do are either polynomial curve fits or exponential curve fits (includes power laws, e.g., $y = ax^b$). If we want to fit a polynomial of order n through our data, we can use the MATLAB's built-in function `polyfit` which already does a least squares curve fit. Therefore, no special effort is required. Now what if we want to fit a non-polynomial function? Two such most commonly used functions are:

1. $y = ae^{bx}$
2. $y = cx^d$

We can convert these exponential curve fits into polynomial curve fits (actually a linear one) by taking log of both sides of the equations, i.e.,

1. $\ln(y) = \ln(a) + bx$ or $\tilde{y} = a_0 + a_1 x$ where $\tilde{y} = \ln(y)$, $a_1 = b$, and $a_0 = \ln(a)$.
2. $\ln(y) = \ln(c) + d\ln(x)$ or $\tilde{y} = a_0 + a_1\tilde{x}$ where $\tilde{y} = \ln(y)$, $a_1 = d$, $a_0 = \ln(c)$ and $\tilde{x} = \ln x$.

Now we can use `polyfit` in both cases with just first order polynomials to determine the unknown constants. The steps involved are the following.

Step-1: Prepare new data: Prepare new data vectors $\tilde{y}$ and $\tilde{x}$, as appropriate, by taking the log of the original data. For example, to fit a curve of the type $y = ae^{bx}$, create `ybar = log(y)` and leave x as it is; while to fit a curve of the type $y = cx^d$, create `ybar = log(y)` and `xbar = log(x)`.

Step-2: Do a linear fit: Use `polyfit` to find the coefficients a_0 and a_1 for a linear curve fit.

Step-3: Plot the curve: From the curve fit coefficients, calculate the values of the original constants (e.g., a, b, etc.). Recompute the values of y at the given x's according to the relationship obtained and plot the curve along with the original data.

Here is an example of one such curve fit. The following table shows the time versus pressure variation readings from a vacuum pump. We will fit a curve, $P(t) = P_0 e^{-t/\tau}$, through the data and determine the unknown constants P_0 and τ.

t	0	0.5	1.0	5.0	10.0	20.0
P	760	625	528	85	14	0.16

By taking log of both sides of the relationship, we have

$$\ln(P) = \ln(P_0) - \frac{t}{\tau}$$

$$\text{or,} \quad \tilde{P} = a_1 t + a_0$$

where $\tilde{P} = \ln(P)$, $a_1 = -1/\tau$, and $a_0 = \ln(P_0)$. Thus, we can easily compute P_0 and τ once we have a_1 and a_0. The following script file shows all the steps involved. The results obtained are shown both on the linear scale and on the log scale in Fig. 5.6.

```
% EXPFIT: Exponential curve fit example
% For the following data for (t,p) fit an exponential curve
%        p = p0 * exp(-t/tau).
% The problem is solved by taking log and then using a linear
% fit (1st order polynomial)

% original data
t=[0 0.5 1 5 10 20];
p=[760 625 528 85 14 0.16];

% Prepare new data for linear fit
tbar = t;                      % no change in t is required
pbar = log(p);

% Fit a 1st order polynomial through (tbar,pbar)
a = polyfit(tbar,pbar,1);     % the output is a = [a1 a0]

% Evaluate constants p0 and tau
p0 = exp(a(2));                % since a(2) = a0 = log(p0)
tau= -1/a(1);                  % since a1 = -1/tau

% (a) Plot the new curve and the data on linear scale
```

```
tnew = linspace(0,20,20);    % create more refined t
pnew = p0*exp(-tnew/tau);    % evaluate p at new t
plot(t,p,'o',tnew,pnew),grid
xlabel('Time (sec)'), ylabel('Pressure (torr)')

% (b) Plot the new curve and the data on semilog scale
lpnew = exp(polyval(a,tnew));
semilogy(t,p,'o',tnew,lpnew),grid
xlabel('Time (sec)'), ylabel('Pressure (torr)')

% Note: you only need one plot, you can select (a) or (b).
```

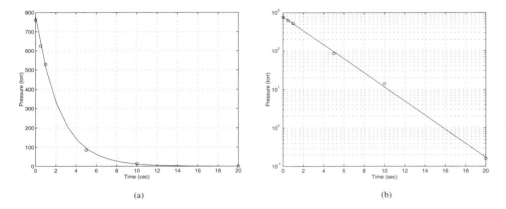

(a)   (b)

Figure 5.6: Exponential curve fit: (a) linear scale plot, (b) semilog scale plot.

There is yet another way to fit a complicated function through your data in the least squares sense. For example, let us say that you have time (t) and displacement (y) data from a spring-mass system experiment and you think that the data should follow

$$y = a_0 \cos(t) + a_1 t \sin(t)$$

Here the unknowns are only a_0 and a_1. The equation is nonlinear in t but *linear-in-the-parameters* a_0 and a_1. Therefore, you can set up a matrix equation using each data point and solve for the unknown coefficients. The matrix equation is

$$\begin{bmatrix} \cos(t_1) & t\sin(t_1) \\ \cos(t_2) & t\sin(t_2) \\ \cos(t_3) & t\sin(t_3) \\ \vdots & \vdots \\ \cos(t_n) & t\sin(t_n) \end{bmatrix} \left\{ \begin{array}{c} a_0 \\ a_1 \end{array} \right\} = \left\{ \begin{array}{c} x_1 \\ x_2 \\ x_3 \\ \vdots \\ x_n \end{array} \right\}$$

Now you can solve for the unknowns a_0 and a_1 simply by typing a = A\x where A is the coefficient matrix and x is the vector containing the measured data x. Typically, A is a rectangular matrix and the matrix equation to be solved is overdetermined (more independent equations than unknowns). This, however, poses no problem, since the backslash operator in MATLAB solves the equation in a least squares sense whenever the matrix is rectangular.

5.2.4   General nonlinear fits

In curve fitting, at times, we need to fit nonlinear equations in which the unknown coefficients appear inside nonlinear functions (as opposed to being linear multipliers to the nonlinear terms). It is not unusual to have double exponential curve fits: $y(x) = C_1 e^{\lambda_1} x + C_2 e^{\lambda_2} x$. Here C_1 and C_2 are linear coefficients but λ_1 and λ_2 are nonlinear coefficients. In MATLAB 6 you can do such curve fits by using the services of `fminsearch` which helps you in finding appropriate values of the nonlinear coefficients by minimizing the error arising from a guess for their values. Type `fitdemo` on the command prompt to see a demo of how you can do such curve fits. However, you are not limited to fitting nonlinear functions of only this form. For example, see problem 4 in the Exercises (it is more like an example with step by step instructions) to find out how to fit a function like $x(t) = C e^{\lambda_1 t} \sin \lambda_2 t$.

5.2.5   Interpolation

Interpolation is the technique of finding a functional relationship between variables such that a given set of discrete values (data points) of the variables satisfy that relationship.[3] Usually, we get a finite set of data points from experiments. When we want to pass a smooth curve through these points or find some intermediate points, we use the technique of interpolation. Interpolation is NOT curve fitting, in that it requires the interpolated curve to pass through all the data points.

In MATLAB 6, you can interpolate your data using *splines* or *Hermite interpolants* on a fly. All you have to do is plot the raw data and use sp ine interpolant or hermite interpolant from the Basic Fitting window options that you can invoke from Tools menu of your Figure window (see Fig. 5.2 on page 123).

In its programming environment, MATLAB provides the following functions to facilitate interpolation:

---

[3]I asked some of my colleagues to define interpolation. One of them gave me this definition, "Reverse engineering an entity of higher dimensionality from information about some entity of lower dimensionality within an identified domain."

interp1: i.e., given y_i at x_i, finds y_j at desired x_j from $y_j = f(x_j)$. Here f is a continuous function that is found from interpolation. It is called one dimensional interpolation because y depends on a single variable x. The calling syntax is

$$\text{ynew = interp1(x,y,xnew,}\textit{method}\text{)}$$

where *method* is an optional argument discussed below after the descriptions of `interp2` and `interp3`.

interp2: Two dimensional data interpolation, i.e., given z_i at (x_i, y_i), finds z_j at desired (x_j, y_j) from $z = f(x, y)$. The function f is found from interpolation. It is called two dimensional interpolation because z depends on a two variables x and y.

$$\text{znew = interp2(x,y,z,xnew,ynew,}\textit{method}\text{)}.$$

interp3: Three dimensional analogue of `interp1`, i.e., given v_i at (x_i, y_i, z_i), finds v_j at desired (x_j, y_j, z_j).

$$\text{vnew = interp3(x,y,z,v,xnew,ynew,znew,}\textit{method}\text{)}.$$

In addition, there is an *n*-dimensional analogue, `interpn`, if you ever need it. As shown above, in each function, you have an option of specifying a *method* of interpolation. The choices for *method* are: *nearest, linear, cubic,* or *spline* The choice of the method dictates the smoothness of the interpolated data. The default method is *linear*. To specify cubic interpolation instead of linear, for example, in `interp1`, use the syntax

$$\text{ynew = interp1(x,y,xnew,'cubic').}$$

The example at the end of this section shows how to use `interp1`. It also shows a comparison of results obtained from different interpolation methods.

There are some other important interpolation functions that are worth mentioning:

Spline: One dimensional interpolation which uses cubic splines to find y_j at desired x_j, given y_i at x_i. Cubic splines fit separate cubic polynomials between successive data points by matching the slopes as well as the curvature of each segment at the given data points. The calling syntax is:

$$\text{ynew = spline(x,y,xnew,}\textit{method}\text{)}$$

There are other variants of the calling syntax. It is also possible to get the coefficients of the interpolated cubic polynomial segments that can be used later. See the on-line help.

interpft: based one dimensional data interpolation. This is similar to `interp1` except that the data is interpolated first by taking the Fourier transform of the given data and then calculating the inverse transform using more data points. This interpolation is especially useful for periodic functions (i.e., if values of y are periodic). See the on-line help.

Example:

There are two simple steps involved in interpolation — providing a list (a vector) of points at which you wish to get interpolated data (this list may include points at which data is already available), and executing the appropriate function (e.g., `interp1`) with the desired choice for the method of interpolation. We illustrate these steps through an example on the x and y data given in the table below.

x	0	0.785	1.570	2.356	3.141	3.927	4.712	5.497	6.283
y	0	0.707	1.000	0.707	0.000	-0.707	-1.000	-0.707	-0.000

Step-1: Generate a vector `xi` containing desired points for interpolation

```
% take equally spaced fifty points.
xi = linspace(0,pi,50);
```

Step-2: Generate data `yi` at `xi`

```
% generate yi at xi with cubic interpolation.
yi = interp1(x,y,xi,'cubic');
```

Here, `'cubic'` is the choice for interpolation scheme. The other schemes we could use are **nearest**, **linear** and **spline**. The data generated by each scheme is shown in Fig. 5.7 along with the original data. The corresponding curves show the smoothness obtained.

Caution: In all interpolation functions, it is required that the input data points in x be monotonic (i.e., either continuously increasing or decreasing).

5.3   Data Analysis and Statistics

Tools menu of Figure window

For performing simple data analysis tasks, such as finding mean, median, standard deviation, etc., MATLAB 6 provides an easy graphical interface that you can activate from the **Tools** menu of the **Figure** window. First, you should plot your data in the form you wish (e.g. scatter plot, line plot, etc). Then, go to the **Figure** window and select **Data Statistics** from the **Tools** pull-down menu. MATLAB shows you the basic stats of your data in a separate window marked **Data Statistics**. You can show any of the statistical measures on your plot by checking the appropriate box (see Fig. 5.8).

However, you are not limited to this simple interface for your statistical needs. Several built-in functions are at your disposal for statistical calculations. These functions are briefly discussed below.

All data analysis functions take both vectors and matrices as arguments. When a vector is given as an argument, it does not matter whether it is a row vector or a column vector. However, when a matrix is used as an argument,

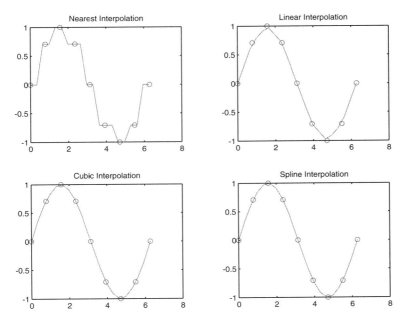

Figure 5.7: 1-D interpolation of data with different schemes: (a) nearest, (b) linear, (c) cubic, and (d) spline.

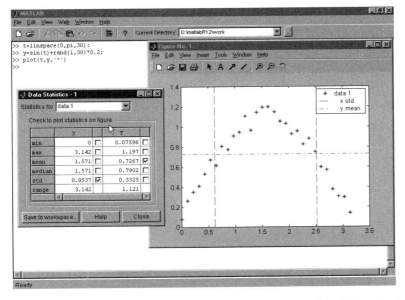

Figure 5.8: Simple data statistics is available to you on the click of a button in MATLAB 6 from the **Tools** menu of the figure window.

the functions operate columnwise on the matrix and output a row vector that contains results of the operation on each column. In the following description of the functions, we will use two arguments in the examples — a row vector x and a matrix A that are given below.

$$x = \begin{bmatrix} 1 & 2 & 3 & 4 & 5 \end{bmatrix}, \qquad A = \begin{bmatrix} 6 & 5 & -2 \\ 7 & 4 & -1 \\ 8 & 3 & 0 \\ 9 & 2 & 1 \\ 10 & 2 & 2 \end{bmatrix}.$$

mean gives the arithmetic mean $\bar{x}$ or the average of the data.
   Example: mean(x) gives 3 while mean(A) results in [8 3.2 0].

median gives the middle value or the arithmetic mean of the two middle values of the data.
   Example: median(x) gives 3 while median(A) gives [8 3 0].

std gives the standard deviation σ based on $n-1$ samples. A flag of value 1 is used as an optional argument (e.g., std(x,1)) to get the standard deviation based on n samples.
   Example: std(x) gives 1.5811, std(x,1) gives 1.4142, and std(A) gives [1.5811 1.3038 1.5811].

max finds the largest value in the data set.
   Example: max(x) gives 5 andmax(A) gives [10 5 2].

min finds the smallest value in the data set.
   Example: min(x) gives 1 andmin(A) gives [6 2 -2].

sum computes the sum Σx_i of the data.
   Example: sum(x) gives 15 while sum(A) gives [40 16 0].

cumsum computes cumulative sum.
   Example: cumsum(x) produces [1 3 6 10 15].

prod computes the product $\prod x_i$ of all data values.
   Example: prod(x) gives 120 and prod(A) gives [30240 240 0].

cumprod computes cumulative product.
   Example: cumprod(x) produces [1 2 6 24 120].

sort sorts the data in ascending order. An optional output argument (e.g., [y,k] = sort(x)) gives the indices of the sorted data values.
   Example: Let z = [22 18 35 44 9]. Then,
   sort(z) results in [9 18 22 35 44],
   [y,j]=sort(z) gives y=[9 18 22 35 44], and j=[5 2 1 3 4], where elements of j are the indices of the data in z that follow ascending order. Thus z = z(j) will results in a sorted z.
   The index vector of the sorted data is typically useful in a situation where you want to sort a few vectors or an entire matrix based on the

sorting order of just one vector or a column. For example, say you have a $m \times n$ matrix B that you want to sort according to the sorted order of the first column of B. You can do this simply with the commands: `[z,j]=sort(B(:,1)); Bnew=sort(B(j,:))`.

Note: As you can see, `sort` sorts data in ascending order. What if you want to sort in descending order? Well, how about sorting the data in ascending order and then `flipping` it upside down with `flipud`. Try `flipud(sort(A))` to get all columns of A arranged in descending order.

`sortrows` sorts the rows of a matrix.

`diff` computes the difference between the successive data points. Thus `y=diff(x)` results in a vector y where $y_i = x_{i+1} - x_i$. The resulting vector is one element shorter than the original vector. `diff` can be used to get approximate numerical derivatives. See the on-line help on `diff`.

`trapz` computes the integral of the data (area under the curve defined by the data) using trapezoidal rule .
Example: `trapz(x)` gives 12 while `trapz(A)` gives `[32 12.5 0]`.

`cumtrapz` computes the cumulative integral of the data. Thus `y=cumtrapz(x)` results in a vector y where $y_i = \Sigma_{t_1}^{t_i} x_i \Delta t_i$ (here the data x_i is assumed to be taken at t_i, i.e., $x_1 = x(t_1)$, $x_2 = x(t_2)$, etc.).
Example: `cumtrapz(x)` results in `[0 1.5 4 7.5 12]`.

In addition to these most commonly used statistical functions, MATLAB also provides functions for correlation and crosscorrelation, covariance, filtering, and convolution. See the on-line help on **datafun**. If you do a lot of statistical analysis, then it may be worth getting the *Statistical Toolbox*.

5.4   Numerical Integration (Quadrature)

Numerical evaluation of the integral $\int f(x)dx$ is called quadrature. Most often, the integrand $f(x)$ is quite complicated and it may not be possible to carry out the integration analytically. In such cases, we resort to numerical integration. However, we can only evaluate definite integrals, i.e., $\int_a^b f(x)dx$, numerically. There are several methods for numerical integration. Consult your favorite book[4] on numerical methods for a discussion of methods, formulas, and algorithms. MATLAB provides the following built-in functions for numerical integration.

For on-line help type:
`help funfun`

---

[4]Some of my favorites: *Applied Numerical Analysis*, by Gerald and Wheatley, Addison Wesley, and *Numerical Recipes*, by Press, Flannery, Teukolski, and Vetterling, Cambridge University Press.

quad: It integrates a specified function over specified limits, based on adaptive Simpson's rule. The adaptive rule seeks to improve accuracy by adaptively selecting the size of the subintervals (as opposed to keeping it constant) within the limits of integration while evaluating the sums that make up the integral.

quadl: It integrates a specified function over specified limits, based on adaptive Lobatto quadrature. This one is more accurate than **quad** but it also uses more function evaluations. It may, however, be more efficient if your integrand is a smooth function. This function is a replacement of **quad8** that existed in MATLAB 5.x. It uses a better and more reliable algorithm.

The general call syntax for both **quad** and **quadl** is as follows:

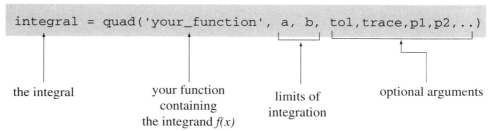

To use **quadl**, you just replace **quad** with **quadl** in the syntax. As shown in the syntax, both functions require you to supply the integrand as a user written function. The optional input argument **tol** specifies absolute tolerance (the default value is 10^{-6}). A nonzero value of the other optional argument, **trace**, shows some intermediate calculations (see on-line help) at each step. The optional arguments **p1**, **p2**, etc., are simply passed on to the user defined function as input arguments in addition to x.

The steps involved in numerical integration using these built-in functions are:

Step-1: Write a function that returns the value of the integrand $f(x)$ given the value of x. Your function should be able to accept the input x as a vector and, correspondingly, produce the value of the integrand (the output) as a vector.

Step-2: Decide which function to use — **quad** or **quadl** (**quad** is faster but less accurate than **quadl**). Find the integral. Decide if you need to change the default tolerance — **tol** $= 10^{-6}$. As a general guideline, if you use **quad** and you are not happy with the answer, use **quadl** before you start fiddling with the tolerance.

Example

Let us compute the following integral

$$\int_{1/2}^{3/2} e^{-x^2} dx.$$

This integral is closely related to the error function, *erf*. In fact,

$$\int_{0}^{x} e^{-x^2} dx = \frac{\sqrt{\pi}}{2} \text{erf}(x).$$

Since MATLAB also provides the error function, `erf`, as a built-in function, we can evaluate our integral in closed form (in terms of the error function) and compare the results from numerical integration. Let us follow the steps outlined above.

Step-1: Here is the function that evaluates the integrand at given x (x is allowed to be a vector).

```
function y = erfcousin(x);
% ERFCOUSIN function to evaluate exp(-x^2).
y = exp(-x.^2)  % the .^ operator is used for vector x
```

Step-2: Let us use `quad` with its simplest syntax:

```
>> y = quad('erfcousin',1/2,3/2)      % here a=1/2, b=3/2

y =

    0.3949
```

The exact result for the integral, up to 10 decimal places, is 0.3949073872. In the example above, we have used the default tolerance. Here is a table showing the results of a few experiments with this integral. We use both `quad` and `quadl` for integration. For `quad`, we tabulate results with different tolerances. In the table, we list the value of the integral, % error, and number of function evaluations,[5] *F-evals* (a measure of computations involved). As you can see, `quadl` gives quite an accurate solution with just the default tolerance. Even `quad` does very well and gives only 0.0000014% error with the default tolerance and just about the same computational effort as `quadl`.

Function	tol	Answer	% Error	F-evals
quad	default	0.3949073929	0.0145×10^{-4}	17
	10^{-7}	0.3949071027	0.0046×10^{-4}	25
	10^{-8}	0.3949073615	0.0002×10^{-4}	33
quadl	default	0.3949073875	0.0008×10^{-4}	18

---

[5]The number of function evaluations can be obtained by including an optional output argument in the call syntax of `quad`: [y,fnevals] = quad('erfcousin', ...).

5.4.1   Double integration

To evaluate integrals of the form

$$\int_{y_{min}}^{y_{max}} \int_{x_{min}}^{x_{max}} f(x,y) \ dx \ dy$$

MATLAB provides a function dblquad (from version 5.2 onwards). The calling syntax for dblquad is:

```
I = dblquad('fxy_fun', xmin, xmax, ymin, ymax, tol, @method)
```

where *tol* and *method* are optional input arguments. The optional argument *tol* specifies tolerance (default is 10^{-6}) as discussed for 1-D integration above, and *method* specifies a choice that the user makes about the method of integration to be used, e.g., quad or quadl. The default method is quad. The user defined integrand function, fxy_fun must be written such that it can accept a vector x and a scalar y while evaluating the integrand.

Since the double integrals are performed by carrying out 1-D integrals in one direction while holding the other variable constant and then repeating the procedure in the second direction, both quad and quadl are valid choices for the *method*. In addition, you could specify another *method* that you have programmed in a function, but your function must have the same input and output arguments, and in the same order, as quad.

Example

Let us compute the following integral

$$I = \int_{-1}^{1} \int_{0}^{2} 1 - 6x^2 y \ dx \ dy.$$

It is fairly simple to verify analytically that $I = 4$. Let us see how dblquad performs on this integral.

```>> F = inline('1-6*x.^2*y');```
```>> I = dblquad(F,0,2,-1,-1)```
 	
```I =```
 	
```   4```
	Create the integrand as an inline function. Note that x is taken as a vector argument. Next, run dblquad with default *tol* and *method*.

Note that we get the exact result just with default tolerance and the default lower order method, quad. You can verify that the higher order method quadl gives the same result by executing the command

```
I = dblquad(F,0,2,-1,1,[],@quadl)
```

Non-rectangular domains

The limits of integration used by `dblquad` are constants (a, b, c and d). Thus the domain of integration is a rectangle. What if the inner limits of integration are variable rather than constant (i.e., the domain of integration is not a rectangle)? It is possible to do a change of variables such that the limits of integration become constant. See a book on numerical methods[6] for a description of the method. Once you convert the limits to constants, you can use `dblquad` to evaluate the integral. Alternatively, you can still use a rectangular domain enclosing the domain of interest and set the value of the integrand to be zero outside the domain of interest using relational operators. Let us take a simple example.

Example: Evaluate
$$\int_0^1 \int_y^1 x^2 e^{xy} \; dx \; dy.$$

The domain of integration here is a triangular region, say D, since x varies from $x = y$ to $x = 1$. We can still use `dblquad` over a rectangular region that contains this triangular region by setting the integrand to be zero outside the triangular region. How do we do that? Well, we can multiply the integrand with another relational expression that returns 0 when $(x, y) \notin D$, and returns 1 if $(x, y) \in D$. This is simple enough; multiply the integrand with `y-x <= 0`. This expression returns 0 when x is to the left of the line $y = x$ and returns 1 when x is to the right of $y = x$. Now we are ready to evaluate the integral.

```
F = inline('x.^2.*exp(x*y) .* (y-x<=0)');   % create the integrand
I = dblquad(F,0,1,0,1)         % integrate over the rectangular domain
```

The given integral happens to be easy enough so that we can integrate analytically and find that $I = e/2 - 1$. The following table shows how `dblquad` performs in this case.

Function	tol	Answer	% Error
`default`	default	0.35917619231246	0.0098
	10^{-8}	0.35914238328561	0.0004
`quadl`	default	0.35914850908091	0.0021
	10^{-8}	0.35914696325471	0.0017

Note that with the same tolerance, the lower order default method (`quad`) performs better in this case than the higher order method `quadl`.

---

[6]For example, *Applied Numerical Analysis* by Gerald and Wheatley, Addition Wesley

For on-line help type:
`help funfun`

5.5 Ordinary Differential Equations

There is a separate suite of Ordinary Differential Equation (ODE) solvers in MATLAB 5.x. The previous versions of MATLAB had just two built-in functions for solution of ODEs — ode23 and ode45. Now there are several additional functions that can also handle stiff equations. Although the various choices now available have increased the versatility of MATLAB's ODE solving capability, ode23 and ode45 remain the workhorses of the suite. In the new version of MATLAB, even ode23 and ode45 have changed and become more versatile. The versatility, however, comes with a cost; the more you want from these functions, the more you have to understand about their complex input structure. At this point, we will not go into such intricate details. We will look at the most straightforward, and perhaps, the most used form for the MATLAB's ODE solvers.

The functions, ode23 and ode45 are implementations of 2nd/3rd-order and 4th/5th-order Runge-Kutta methods, respectively. Solving most ODEs using these functions in their simplest form (without any optional arguments) involves the following four steps:

1. **Write the differential equation(s) as a set of first-order ODEs.** For ODEs of order ≥ 2, this step involves introducing new variables and recasting the original equation(s) in terms of 1st-order ODEs in the new variables. Basically, you need the equation in the vector form $\dot{\mathbf{x}} = \mathbf{f}(\mathbf{x},t)$, where $\mathbf{x} = [x_1 \quad x_2 \quad \ldots \quad x_n]^T$. In expanded form, the equation is:

$$\left\{ \begin{array}{c} \dot{x}_1 \\ \dot{x}_2 \\ \vdots \\ \dot{x}_n \end{array} \right\} = \left\{ \begin{array}{c} f_1(x_1, x_2, \ldots, x_n, t) \\ f_2(x_1, x_2, \ldots, x_n, t) \\ \vdots \\ f_n(x_1, x_2, \ldots, x_n, t) \end{array} \right\}$$

2. **Write a function to compute the state derivative.** The state derivative $\dot{\mathbf{x}}$ is the vector of derivatives $\dot{x}_1, \dot{x}_2, \ldots, \dot{x}_n$. Therefore, you have to write a function that computes $f_1, f_2, \ldots, f_n$, given the input $(\mathbf{x}, t)$ where $\mathbf{x}$ is a column vector, that is, $\mathbf{x} = [x_1 \quad x_2 \quad \ldots \quad x_n]^T$. Your function must return the state derivative $\dot{\mathbf{x}}$ as a column vector.

3. **Use the built-in ODE solvers ode23 or ode45 to solve the equations.** Your function written in Step-2 is used as an input to ode23 or ode45. The syntax of use of ode23 is shown below. To use ode45 just replace 'ode23' with 'ode45'.

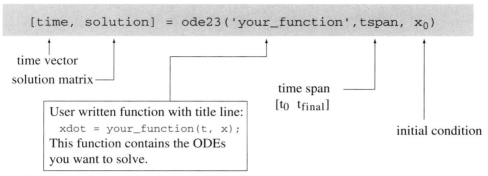

4. **Extract the desired variables from the output and interpret the results.** For a system of n equations, the output matrix `solution` contains n columns. You need to understand which column corresponds to which variable in order to extract the correct column, if you want to plot a variable with respect to, say, the independent variable `time`.

Here are two examples.

5.5.1   Example–1: A first-order linear ODE

Solve the first-order linear differential equation

$$\frac{dx}{dt} = x + t \tag{5.3}$$

with the initial condition

$$x(0) = 0.$$

Step-1: Write the equation(s) as a system of first-order equations: The given equation is already a first-order equation. No change is required.

$$\dot{x} = x + t.$$

Step-2: Write a function to compute the new derivatives: The function should return $\dot{x}$ given x and t. Here is the function:

```
function xdot = simpode(t,x);
% SIMPODE: computes xdot = x+t.
% call syntax:  xdot = simpode(t,x);
xdot = x + t;
```

Write and save it as an M-file named **simpode.m**.

Step-3: Use ode23 to compute the solution: The commands as typed in the command window are shown below. These commands could instead be part of a script file or even another MATLAB function. Note that we have not used the optional arguments tol or trace.

```
>> tspan = [0  2]; x0 = 0;   % specify values of tspan and x0
>> [t,x] = ode23('simpode', tspan, x0);   % now execute ode23
```

Step-4: Extract and interpret results: The output variables t and x contain results— t is a vector containing all discrete points of time at which the solution was obtained, and x contains the values of the variable x at those instances of time. Let us see the solution graphically:

```
>> plot(t,x)                  % plot t vs x and label axes
>> xlabel('t'), ylabel('x')   % put labels
```

The plot generated by above commands is shown in Figure 5.9.

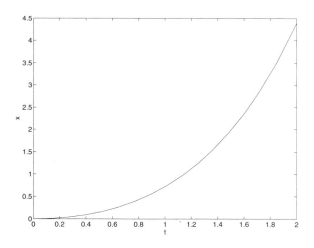

Figure 5.9: Numerical solution of the equation using ode23.

5.5.2   Example–2: A second-order nonlinear ODE

Solve the equation of motion of a nonlinear pendulum

$$\ddot{\theta} + \omega^2 \sin\theta = 0 \quad \Rightarrow \quad \ddot{\theta} = -\omega^2 \sin\theta \tag{5.4}$$

with the initial conditions

$$\theta(0) = 1, \quad \dot{\theta}(0) = 0.$$

Step-1: Write the equation(s) as a system of first-order equations:
The given equation is a second-order ODE. To recast it as a system of two first-order equations (an nth-order equation reduces to a set of n 1st-order equations), let us introduce two new variables.
Let $z_1 = \theta$ and $z_2 = \dot{\theta}$. Then $\dot{z}_1 = \dot{\theta} = z_2$ and $\dot{z}_2 = \ddot{\theta} = -\omega^2 \sin(z_1)$.
Now eqn. (5.4) may be written in vector form as

$$\left\{ \begin{array}{c} \dot{z}_1 \\ \dot{z}_2 \end{array} \right\} = \left\{ \begin{array}{c} z_2 \\ -\omega^2 \sin(z_1) \end{array} \right\}.$$

$$\dot{\mathbf{z}} = \mathbf{f}(\mathbf{z})$$

components each:

$$\mathbf{z} = \left\{ \begin{array}{c} z_1 \\ z_2 \end{array} \right\}, \qquad \dot{\mathbf{z}} = \left\{ \begin{array}{c} \dot{z}_1 \\ \dot{z}_2 \end{array} \right\} = \left\{ \begin{array}{c} z_2 \\ -\omega^2 \sin z_1 \end{array} \right\} = \mathbf{f}(\mathbf{z}).$$

This is a special case of $\dot{\mathbf{z}} = \mathbf{f}(t, \mathbf{z})$ where $\mathbf{f}$ does not depend on t.

Step-2: Write a function to compute the new state derivative: We need to write a function that, given the scalar time t and vector $\mathbf{z}$ as input, returns the time derivative vector $\dot{\mathbf{z}}$ as output. In addition, the state derivative vector $\dot{\mathbf{z}}$ must be a column vector. Here is a function that serves the purpose.

```
function zdot = pend(t,z);
% Call syntax: zdot = pend(t,z);
% Inputs are:  t = time
%              z = [z(1); z(2)] = [theta; thetadot]
% Output is :  zdot = [z(2); -w^2 sin z(1)]
wsq = 1.56;   % specify a value of w^2
zdot = [z(2); -wsq*sin(z(1))];
```

Note that $z(1)$ and $z(2)$ refer to the first and second elements of vector $\mathbf{z}$. Do not forget to save the function 'pend' as an M-file to make it accessible to MATLAB.

Step-3: Use ode23 or ode45 for solution: Now, let us write a script file that solves the system and plots the results. Remember that the output $\mathbf{z}$ contains two columns: z_1, which is actually θ, and z_2, which is $\dot{\theta}$. Here is a script file that executes ode23, extracts the displacement and velocity in vectors x and y, and plots them against time as well as in the phase plane.

```
tspan = [0 20]; z0 = [1;0];  % assign values to tspan, z0
[t,z] = ode23('pend',tspan,z0);    % run ode23
x = z(:,1); y = z(:,2);      % x=1st column of z, y=2nd column
```

```
plot(t,x,t,y)                    % plot t vs x and t vs y
xlabel('t'), ylabel('x and y')
figure(2)                        % open a new figure window
plot(x,y)                        % plot phase portrait
xlabel('Displacement'), ylabel('Velocity')
title('Phase Plane of a Non-linear Pendulum') % put a title
```

Step-4: Extract and interpret results: The desired variables have already been extracted and plotted by the script file in Step-3. The plots obtained are shown in Fig. 5.10 and Fig. 5.11.

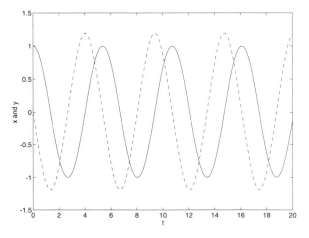

Figure 5.10: Displacement and velocity vs time plot of the pendulum.

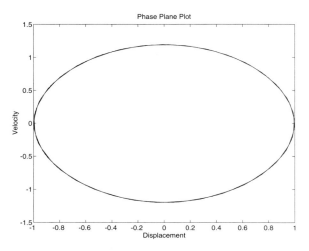

Figure 5.11: Displacement vs velocity plot of the pendulum.

5.5.3 ode23 versus ode45

For solving most initial value problems, you use either ode23 or ode45. Which one should you choose? In general, ode23 is quicker but less accurate than ode45. However, the actual performance also depends on the problem. As the following table shows, ode45 is perhaps a better choice. We solved the two simple ODEs discussed above in Sections 5.5.1 and 5.5.2 with these functions and noted down the number of successful steps (*Good steps*), number of failed steps (*Bad steps*), number of function evaluations (*f-evals*), and number of floating point operations (*flops*). Solutions were obtained with different *relative tolerances*. What does this tolerance mean? How do you set it? We discuss this in the following sections (Sections 5.5.4 and 5.5.5).

RelTol	Solver	Good steps	Bad Steps	f-evals
Solve $\dot{x} = x + t$,		$x(0) = 0$,	$0 \leq t \leq 2$	
default	ode23	17	3	61
default	ode45	10	0	61
Solve $\ddot{\theta} = -\omega^2 \sin\theta$,		$\theta(0) = 1, \dot{\theta}(0) = 0$,	$0 \leq t \leq 20$	
default	ode23	114	13	382
default	ode45	33	0	199
10^{-4}	ode23	251	15	799
10^{-4}	ode45	49	1	301
10^{-5}	ode23	528	0	1585
10^{-5}	ode45	80	10	541
10^{-6}	ode23	805	0	2416
10^{-6}	ode45	104	0	625

5.5.4 Specifying tolerance

Many functions in the funfun category provide the user with a choice of specifying a *tolerance* as an optional input argument. A tolerance is a small positive number that governs the error of some appropriate kind in the computations of that function. For example, the ODE solvers ode23 or ode45 use the tolerance in computing the step size based on the error estimate of the computed solution at the current step. So what value of the tolerance should you specify? To make your decision even harder, there are two of them — relative tolerance and absolute tolerance.

The relative tolerance, in general, controls the number of correct digits in the solution of a component. Thus, you want the solution to have k digit accuracy, you should set the relative tolerance to 10^{-k}. This is because the error estimated in the solution of a component y_i is compared with the value *Relative Tolerance* $\times |y_i|$. In fact, MATLAB uses a combination of both the relative tolerance and the absolute tolerance. For example, in ODE solvers, the estimated local error e_i in the solution component y_i at any step is required to satisfy the relationship:

$$e_i \leq \max(RelTol \times |y_i|, AbsTol).$$

The absolute tolerance specifies the threshold level for any component of the solution below which the computed values of the component have no guarantee of any accuracy. From the error relationship shown above, it should be clear that specifying a very low value of relative tolerance, below the absolute tolerance, may not lead to more accurate solutions.

The absolute tolerance can also be specified as a vector in which case it specifies the threshold values of each solution component separately. The value of a solution component below the corresponding absolute tolerance is almost meaningless.

5.5.5   The ODE Suite

The ODE Suite of MATLAB 5.x consists of several other solvers (mostly for stiff ODEs) and utility functions. Even the old faithfuls, `ode23` and `ode45`, have been rewritten for better performance. The following list shows some of the functions and utilities that are available in MATLAB 5.x. You should see the on-line help on these functions before using them.

ode45	non-stiff solver based on 4/5 order Runge-Kutta method
ode15s	stiff solver based on a variable order method
ode23	non-stiff solver based on 2/3 order Runge-Kutta method
ode113	non-stiff solver based on variable order Adams-Bashforth-Moulton methods
ode23t	stiff solver for moderately stiff equations, based on trapezoidal rule
ode23s	stiff solver based on a low order Numerical Differentiation Formula (NDF)
ode23tb	stiff solver based on a low order method

Utility functions

odefile	a help file to guide you with the syntax of your ODE function
odeset	a function that sets various `options` for the solvers
odeget	a function that gets various `options` parameters
odeplot	a function that can be specified in `options` for time series plots
odephas2	a function that can be specified in `options` for 2D phase plots

Since there are so many choices for solvers, there is likely to be some confusion about which one to choose. If you do not have stiff equations, your choices are limited to `ode23`, `ode45`, and `ode113`. A general rule of thumb is that `ode45` will give you satisfactory results for most problems, so try it first. If you suspect the equations to be stiff, try `ode15s` first from the list of stiff solvers. If that doesn't work, learn about stiffness, understand your equations better, and find out more about the stiff solvers in the ODE suite before you start trying them out.

One nice feature of the entire ODE Suite is that all solvers use the same syntax. We discussed the simplest form of the syntax in the previous section. Now let us look at the full syntax of these functions:

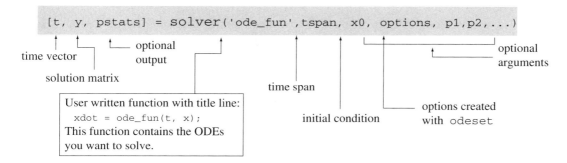

There are three new optional arguments here — one in the output, `stats`, and two in the input, `options` and `p1, p2, ···, pn`. Here is a brief description of these arguments.

`pstats`: It is a six element long vector that lists the performance statistics of the solvers. The first three elements are (i) numbers of successful steps, (ii) number of failed steps, and (iii) number of function evaluations. The next three elements are related to performance indices of stiff solvers.

`options`: function `odeset` with the command:

$$\text{options} = \text{odeset}('name1', \text{value1}, 'name2', \text{value2}, \cdots)$$

where *name* refers to the name of the optional argument that you want to set and `value` refers to the corresponding value of the argument. The list of the arguments that you can set with `odeset` includes relative tolerance (`reltol`), absolute tolerance (`abstol`), choice of output function (`outputfcn`), performance statistics (`stats`), etc. There are many more optional arguments. See the on-line help on `odeset`. The most commonly used arguments, perhaps, are `reltol`, `abstol`, `stats`, and `outputfcn`. When `stats` is `'on'`, the solver performance statistics is displayed at the end. It is an alternative[7] to specifying the optional

---

[7]Of course, if you specify the optional output argument, the statistics is saved in that variable whereas with `'stats' 'on'` in the options, you see the statistics but cannot have access to them later.

output argument `pstats` discussed above. The argument `outputfcn` can be assigned any of the output functions, `'odeplot'`, `'odephas2'`, `'odephas3'`, or `'odeprint'` to generate time series plot, phase plot in 2-D or 3-D, or show the computed solution on the screen, as the solution proceeds.

As an example, suppose we want to set the relative tolerance to 10^{-6}, absolute tolerance to 10^{-8}, see the performance statistics, and have MATLAB plot the time series of the solution at each time step as the solution is computed, we set the `options` structure as follows:

```
options = odeset( 'reltol',1e-6,'abstol',1e-8,...
                  'stats','on', 'outputfcn','odeplot');
```

`p1, p2, ..., pn`: These are optional arguments that are directly passed on to the user written ODE function. In the previous versions of MATLAB, the only arguments that the solvers could pass on to the ODE function were (`t, x`). Thus, if the user written function needed other variables to be passed, they had to be either hardcoded or passed on through `global` declaration. To find out how to use these variables in the ODE function and how to write the list of input variables to incorporate these optional variables, please see the on-line help on `odefile`.

5.5.6 Event location

In solving initial value problems, usually the termination condition is specified in terms of the independent variable. As shown in the previous two examples, the solution stops when the independent variable t reaches a prescribed value t_{final} (specified as the second element of `tspan`). Thus, we obtain the solution for a certain time span. Sometimes, however, we need to stop the solution at a specified value of the dependent variable and we do not know when (at what value of t) the solution will reach there. For example, say, we want to solve a projectile problem. We write the equations of motion (two ODEs for $\ddot{x}$ and $\ddot{y}$), and we would like to solve the equations to find *when* the projectile hits the target. Since we do not know t_{final} apriori, what value of t_{final} should we specify in `tspan`? One thing we know is that the solution should stop when the target (some specified location (x_{target} or y_{target}) is hit, that is, when x or y reach a particular value. When the dependent variables (here $x(t)$ or $y(t)$) reach some specified value, we call that an *event* has occurred, and the problem of finding time of the event is called *event location*. Event location problems are solved by following the solution till the solution *crosses* the event, then backtracking to the time just before the event, and then taking smaller and suitably computed time step to land exactly at the event. There are several strategies and algorithms for event location. In the literature, the event location problem is also referred to as *integrating across discontinuities.*

Fortunately, the ODE solvers in MATLAB 5.x have built-in ability to solve event location problems. Let us take a simple example — a projectile is thrown with initial speed v_o at an angle θ. We want to find out (i) the instant when the projectile hits the ground, (ii) the range, and (iii) the trajectory of the projectile. The ODEs governing the motion of the projectile are

$$\ddot{x} = 0$$
$$\ddot{y} = -g$$

The initial conditions are $x(0) = y(0) = 0$, $\dot{x}(0) = v_0 \cos\theta$, $\dot{y}(0) = v_0 \sin\theta$. The equations of motion are to be integrated till $y = 0$ (the projectile hits the ground).

Before we write a function to code our equations, we need to convert them into a set of first order equations:

Let $x = x_1, \dot{x} = x_2, y = x_3$, and $\dot{y} = x_4$ then $\dot{x}_1 = x_2, \dot{x}_2(= \ddot{x}) = 0, \dot{x}_3 = x_4$, and $\dot{x}_4(= \ddot{y}) = -g$. In vector form:

$$\left\{ \begin{array}{c} \dot{x}_1 \\ \dot{x}_2 \\ \dot{x}_3 \\ \dot{x}_4 \end{array} \right\} = \left\{ \begin{array}{c} x_2 \\ 0 \\ x_4 \\ -g \end{array} \right\} \tag{5.5}$$

Now, equation (5.5) is ready for coding in a function.

Event specification

In this example, the event takes place at $y = 0$. But if we somehow instruct the solver to stop when $y = 0$, our projectile will never take off because the solver will stop right at the start, since $y(o) = 0$. Therefore, we have to make the event detection more robust by adding some more condition(s). In this example, we can use the fact that the vertical component of the velocity will be negative when the projectile hits the ground, i.e., $\dot{y} < 0$. Thus, the conditions for the event detection are

$$y = 0, \qquad \dot{y} < 0.$$

Once the event is detected, we have to take another decision — what to do next. We could stop the solution (as we desire in the present example) or we could simply note the event and continue with or without changing some variables (for example, a ball bouncing off a wall would require the detection of collision with the wall, application of collision laws to determine the reversed initial velocity and then continuation of the solution). If the solution has to stop at the event, we will call the event *terminal*, else *non-terminal*. Our event is terminal. This completes the specification of the event.

Solution with event detection

We will now show how to solve the projectile problem using MATLAB's solver ode45 such that the solver automatically detects the collision of the projectile with the ground and halts the program. The steps involved are:

Step-1: Tell the solver to watch out for *events*: We need to tell the solver that we want it to look for *events* that we are going to specify. This is done by setting the 'events' flag 'on' in the options with the function odeset:

$$\text{options = odeset('events','on');}$$

Of course, when we run the solver (see step-3 below), we will use options as an optional input argument.

Step-2: Write your function to give *event* information: We need to write the ODE function with some additional features now. The MATLAB solver is going to call this function with an extra input variable, flag, that tells the ODE function what kind of output is desired. When the solver calls the ODE function with the *flag* set to events, it looks for three output quantities from the ODE function:

1. value: a vector of values of those variables for which a zero crossing defines the *event*. In our example, the event (collision with the ground) is defined by $y = 0$, i.e., a zero crossing of the y-value. Thus, in our case, value = x(3) (the value of y).

2. isterminal: a vector of 1's and 0's specifying whether occurrence of an event is terminal (stops the solution) or not. For each element in the vector value, a 0 or a 1 is required in the vector isterminal. In our example, we have only one element in value (x(3) or y), therefore, we need just one element in isterminal. Since, we would like to stop the program at $y = 0$, the zero crossing of y is terminal. Therefore we set, isterminal = 1.

3. direction: a vector of the same length as value, specifying the direction of zero crossing for each element in value: a -1 for negative value, a 1 for positive value, and a 0 for *I-don't-care* value. In our example, direction = -1 will imply that the zero crossing of y is a valid event only when the value of y is decreasing (y crosses zero with $\dot{y} < 0$). As you can see, specification of direction prevents the solver from aborting the solution at the outset when $y = 0$ but $\dot{y} \not< 0$.

Of course, your function should provide the event information, in terms of these three vectors as output, but only when asked. Thus, your function should check the value of the `flag` and output the three vectors if the `flag` is set to `events`, otherwise it should output the usual derivative vector `xdot`. We can implement this conditional output using `switch` (you can do it using `if-elseif` construction too) as shown in the example function below.

```
function [value,isterminal,dircn] = proj(t,z,flag);
% PROJ: ODE for projectile motion with event detection
g = 9.81;                      % specify constant g
if nargin<3 | isempty(flag) % if no flag or empty flag
   value = [z(2); 0; z(4); -g];
else
   switch flag                 % see Section 4.3.4 for 'switch'
     case 'events'
       value = z(3);           % 'value' is zero crossing for y
       isterminal = 1;         % 'isterminal': y=0 is terminal
       dircn = -1;             % 'direction': ydot < 0
     otherwise
       error('function not programmed for this event');
   end                         % end of switch
end                            % end of if
```

Step-3: Run the solver with appropriate input and output: Clearly, we need to include `options` in the input list. In the output list, we have the choice of including some extra variables:

`te`	times at which the specified events occur
`ze`	solutions (values of state variables) at `te`
`ie`	indices of events that occurred at `te`

So, now let us solve the projectile problem. Our ODE function `proj` is ready. Here is a script file that specifies the required input variables and executes `ode45` with appropriate output variables. The trajectory is shown in Fig. 5.12 along with some information about range and time of flight.

```
% RUNPROJ: a script file to run the projectile example
tspan=[0 2];                   % specify time span, [t0 tfinal]
v0=5; theta = pi/4;            % theta must be in radians
z=[0; v0*cos(theta); 0; v0*sin(theta)];% specify initial conditions
options = odeset('events', 'on');
                               % set the events switch on in options
[t,z,te,ze,ie] = ode45('proj',tspan,z0,options);
                               % run ode45 with optional arguments
x = z(:,1);   y = z(:,3);      % separate out x and y
plot(x,y), axis('equal')       % plot trajectory
```

```
xlabel('x'), ylabel('y'),      % label axes
title('Projectile Trajectory')% write a title
info = ['Range (x-value) = ', num2str(ze(1)), ' m';
        'Time of flight  = ', num2str(te), ' s'];
text([1;1],[0;-.1], info)      % print Range and Time on the plot
```

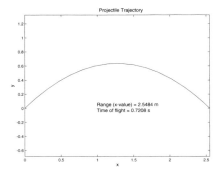

Figure 5.12: Trajectory of a projectile obtained by integrating equations of motion with **ode45** and using the event detection facility of the solver.

5.6   Nonlinear Algebraic Equations

The MATLAB function **fzero** solves nonlinear equations involving one variable (see also function **solve** in Section 8.1 if you have the Symbolic Math Toolbox). You can use **fzero** by proceeding with the following three steps.

1. **Write the equation in the standard form:**

$$f(x) = 0.$$

   This step usually involves trivial rearrangement of the given equation. In this form, solving the equation and *finding a zero of* $f(x)$ are equivalent.

2. **Write a function that computes** $f(x)$**:** The function should return the value of $f(x)$ at any given x.

3. **Use the built-in function fzero to find the solution:** **fzero** requires an initial guess and returns the value of x closest to the guess at which $f(x)$ is zero. The function written in Step-2 is used as an input to the function **fzero**. The call syntax of **fzero** is:

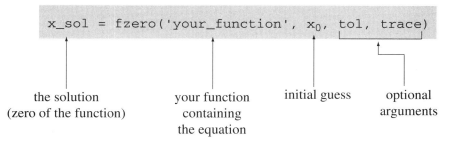

the solution
(zero of the function)

your function
containing
the equation

initial guess

optional
arguments

Example: A transcendental equation

Solve the following transcendental equation:

$$\sin x = e^x - 5.$$

Step-1: Write the equation in standard form: Rearrange the equation
as: $\sin(x) - e^x + 5 = 0 \Rightarrow f(x) = \sin(x) - e^x + 5.$

Step-2: Write a function that computes $f(x)$**:** This is easy enough:

```
function f = transf(x);
% TRANSF: computes f(x) = sin(x)-exp(x)+5.
% call syntax:  f = transf(x);
f = sin(x) - exp(x) + 5;
```

Write and save the function as an M-file named **transf.m**.

Step-3: Use `fzero` **to find the solution:** The commands as typed in the
command window are shown below. The result obtained is also shown.
Note that we have not used the *optional arguments* `tol` or `trace`.

```
>> x = fzero('transf',1)          % initial guess x0=1.

x =

    1.7878
```

To check the result, we can plug the value back into the equation or
plot $f(x)$ and see if the answer looks right. See the plot of the function
in Fig. 5.13.

You can also find the zeros of a polynomial equation (e.g., $x^5 - 3x^3 + x^2 - 9 = 0$) with `fzero`. However, `fzero` locates the root closest to the initial
guess; it does not give all roots. To find all roots of a polynomial equation,
use the built-in function `roots`. Finding multiple roots of a non-polynomial
or finding roots of functions of several variables is a more advanced problem.

We need to mention some new functions that have been added in MATLAB 6
for facilitating optimization calculations. In particular, `fminbnd` and `fminsearch`
have been added to find the unconstrained minimum of a nonlinear function.
See on-line help for more details.

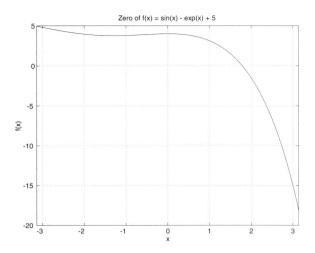

Figure 5.13: **fzero** locates the zero of this function at $x = 1.7878$.

5.7   Advanced Topics

The topics covered in the preceding sections of this chapter are far from being exhaustive in what you can do readily with MATLAB. These topics have been selected carefully to introduce you to various applications that you are likely to use frequently in your work. Once you gain a little bit of experience and some confidence, you can explore most of the advanced features of the functions introduced above as well as several functions for more complex applications on your own, with the help of on-line documentation.

MATLAB 6 provides some new functions for solving two-point boundary value problems, simple partial differential equations, and nonlinear function minimization problems. In particular, we mention the following functions.

bvp4c: solves two-point boundary value problem defined by a set of ODEs of the form $y' = f(x, y)$, and its boundary conditions $y(a)$ and $y(b)$ over the interval $[a, b]$. The user has to write two functions — one that specifies the equations and the other that specifies the boundary conditions. User can set several options for initiating the solutions. To see an example, try executing the built-in example **twobvp** and learn how to program your BVP by following this example (to see the functions required for **twobvp**, type **type twoode.m** and **type twobc.m**).

pdepe solves simple parabolic and elliptic partial differential equations of a single dependent variable. This is a rather restricted utility function. However, for those who need to solve PDEs frequently, there is the *Partial Differential Equation Toolbox.*

EXERCISES

1. **Linear algebraic equations:** Find the solution of the following set of linear algebraic equations as advised below.

$$\begin{aligned}
x + 2y + 3z &= 1 \\
3x + 3y + 4z &= 1 \\
2x + 3y + 3z &= 2.
\end{aligned}$$

   - Write the equation in matrix form and solve for $\mathbf{x} = [x\ y\ z]^T$ using the left division \.
   - Find the solution again using the function `rref` on the augmented matrix.
   - Can you use the LU decomposition to find the solution? [Hint: Since $[\mathbf{LU}]\mathbf{x} = \mathbf{b}$, let $[\mathbf{U}]\mathbf{x} = \mathbf{y}$, so that $[\mathbf{L}]\mathbf{y} = \mathbf{b}$. Now, first solve for $\mathbf{y}$ and then for $\mathbf{x}$.]

2. **Eigenvalues and eigenvectors:** Consider the following matrix.

$$A = \begin{bmatrix} 3 & -3 & 4 \\ 2 & -3 & 4 \\ 0 & -1 & 1 \end{bmatrix}$$

   - Find the eigenvalues and eigenvectors of A.
   - Show, by computation, that the eigenvalues of A^2 are square of the eigenvalues of A.
   - Compute the square of the eigenvalues of A^2. You have now obtained the eigenvalues of A^4. From these eigenvalues, can you guess the structure of A^4?
   - Compute A^4. Can you compute A^{-1} without using the `inv` function?

3. **Linear and quadratic curve fits:** The following data is given to you.

$$\begin{bmatrix} x \\ y \end{bmatrix} = \begin{bmatrix} 0 & 0.15 & 0.2 & 0.30 & 0.4 & 0.52 & 0.6 & 0.70 & 0.8 & 0.90 & 1 \\ 0 & 1.61 & 2.2 & 3.45 & 4.8 & 6.55 & 7.8 & 9.45 & 11.2 & 13.05 & 15 \end{bmatrix}$$

   - Enter vectors x and y and plot the raw data using `plot(x,y,'o')`.
   - Click on the figure window. Select **Basic Fitting** from the **Tools** menu of the figure window. Once you get the **Basic Fitting** window, check the box for **linear** fit. In addition, check the boxes for **Show equations**, **Plot residuals** (select a *scatter* plot rather than a *bar* plot option), and **Show norm of residuals**.
   - Now, do a quadratic fit also by checking the **quadratic** fit box and all other boxes as you did for the linear case.
   - Compare the two fits. Which fit is better? There isn't really a competition, is there?

4. **A nonlinear curve fit:** Enter the following experimental data in MATLAB workspace.

```
t = [0 1.40 2.79 4.19 5.58 6.98 8.38 9.77 11.17 12.57];
x = [0 1.49 0.399 -0.75 -0.42 0.32 0.32 -0.10 -0.21 0];
```

You are told that this data comes from measuring the displacement x of a damped oscillator at time instants t. The response x is, therefore, expected to have the following form:

$$x = Ce^{-\lambda_1 t} \sin(\lambda_2 t)$$

where the constants C, λ_1, and λ_2 are to be determined from the experimental data. Thus, your job is to fit a curve of the given form to the data and find the constants that give the best fit.

- First, see a built-in demo for nonlinear curve fits by typing `fitdemo` (only MATLAB 6).
- Write a function that computes x given some initial guess of the values of the constants C, λ_1 and λ_2 as follows.

  ```
  function er_norm = expsin(constants,t_data,x_data);
  % given lambda, compute norm of the error in the fitted curve
  t = t_data'; x = x_data' ;   % make sure t & x are columns
  C = constants(1); l1 = constants(2); l2 = constants(3);
  xnew = C * exp(-l1 * t) .* sin(l2 *t);   % evaluate your x
  er_norm = norm(xnew - x) % compare with data and compute error
  ```

- Now you can find the best **constants** iteratively by minimizing the norm of the error with the function **fminsearch** as follows.

  ```
  constants = [1 0.1 0.8];   % initial guess of [C, l1, l2]
  constants = fminsearch('expsin',constants,[],t,x);
  ```

- Now that you have the *best fit* constants, evaluate your formula at the given t and plot the fitted curve along with the given data. That's it.

5. **Data statistics:**   MATLAB 6 provides a graphical tool for finding simple statistical measures on data. Consider the data given in Problem 3.

   - Plot the raw data with the marker `'o'`.
   - Select **Data Statistics** from the **Tools** pull-down menu of the figure window. A new window appears that shows the basic statistical measures of the data set.
   - To include any of these measures in your plot, simply check the appropriate box.

6. **Length of a curve through quadrature:** The length of a parametric curve defined by $x(t)$ and $y(t)$ over $a \leq t \leq b$ is given by the integral $\int_a^b \sqrt{(x')^2 + (y')^2} \, dt$ where $x' = dx/dt$ and $y' = dy/dt$. Find the length of a hypocycloid defined by $x(\theta) = a\cos^3\theta$, $y(\theta) = a\sin^3\theta$, $0 \leq \theta \leq \tau/2$. Take $a = 1$.

7. **A second order linear ODE, the phenomenon of beats and resonance:**   Consider the following simple linear ODE of second order.

   $$\ddot{x} + x = F_0 \cos\omega t.$$

   - Convert the given equation to a set of first order ODEs and program the set of equations in a function to be used for numerical solution with `ode45`.

- Set $F_0 = 0$, and solve the system of equations with the initial conditions $x(0) = 0$, $\dot{x}(0) = 1$. By plotting the solution, make sure that you get a periodic solution. Can you find the period of this solution?
- Let $F_0 = 1$ and $\omega = 0.9$. Starting with zero initial conditions ($x(0) = \dot{x}(0) = 0$), find the solution, x and $\dot{x}$ for $0 \leq t \leq 70$. Plot x against t. How is the amplitude of the rapidly oscillating solution modulated? You should see the beats phenomenon.
- Let $F_0 = 1$ and $\omega = 1$. Again starting with zero initial conditions, find the solution for $0 \leq t \leq 40$. How does the amplitude of the solution, x, change with time in this case? This is the phenomenon of resonance.
- Experiment with the solution of the system by adding a damping term, say $\zeta\dot{x}$, on the left hand side of the equation.

8. **Nonlinear ODEs, phase plots, and limit cycles:** Consider the following set of first order, coupled, nonlinear ODEs.

$$\dot{x} = x + y - x(x^2 + y^2) \qquad (5.6)$$
$$\dot{y} = -x + y - y(x^2 + y^2) \qquad (5.7)$$

- Solve this set of equations with the initial conditions $x(0) = 2$ and $y(0) = 2$ over the time interval $0 \leq t \leq 20$. Plot x vs t and y vs t in two different figures. Use `hold on` to keep the plots and graph subsequent solutions as overlay plots. Save the solutions x and y in, say `x1` and `y1`, for later use.
- Solve the given set of equations for the following set of initial conditions and plot the solutions as above (along with the old solutions): (i) $(x_0, y_0) = (1, 0)$, (ii) $(x_0, y_0) = (0.1, 0)$, and (iii) $(x_0, y_0) = (0, 1)$. Save the solutions for each case for later use. Do the plots suggest a particular long time behavior of the system? You should see both $x(t)$ and $y(t)$ settling down to the same periodic motion for all initial conditions.
- The ultimate periodic solution that all solutions seem to get attracted to is called a *limit cycle*. It is much easier to visualize this cycle in the phase plane, that is, the plane of x and y. Plot x vs y for all solutions (you have saved the data from each run) in a single figure. This plot is called *phase plot*. You should clearly see the limit cycle in this plot. (Use `axis('square')` to make the periodic orbit a circular orbit.)
- Try solutions with some other initial conditions.
- **Challenge:** Can you write a script file that will take the initial conditions from your mouse click in the phase plane, solve the equations, and draw the phase plane orbit as many times as you wish? [Hint: you can use `ginput` for taking initial conditions from the mouse click.]

Note: The same system of ODEs given above reduces to very simple and separable equations in polar coordinates:

$$\dot{r} = r(1 - r^2), \quad \dot{\theta} = -1$$

Now, you can even solve the equations analytically and get an expression for the limit cycle as $t \to \infty$.

6. *Graphics*

MATLAB includes good tools for visualization. Basic 2-D plots, fancy 3-D graphics with lighting and color-maps, complete user-control of the graphics objects through *Handle Graphics*, tools for design of sophisticated graphics user-interface, and animation are now part of MATLAB. What is special about MATLAB's graphics facility is its ease of use and expandability. Commands for most garden-variety plotting are simple, easy to use, and intuitive. If you are not satisfied with what you get, you can control and manipulate virtually everything in the graphics window. This, however, requires an understanding of the Handle Graphics, a system of low-level functions to manipulate graphics objects. In this section we take you through the main features of the MATLAB's graphics facilities.

6.1   Basic 2-D Plots

For on-line help type:
`help graph2d`

The most basic and perhaps the most useful command for producing a simple 2-D plot is

$$\texttt{plot}(\textit{xvalues, yvalues, 'style-option'})$$

where *xvalues* and *yvalues* are vectors containing the x- and y-coordinates of points on the graph and the *style-option* is an optional argument that specifies the color, the line style (e.g. solid, dashed, dotted, etc.), and the point-marker style (e.g. o, +, *, etc.). All the three options can be specified together as the *style-option* in the general form:

color_linestyle_markerstyle

The two vectors *xvalues* and *yvalues* MUST have the same length. Unequal length of the two vectors is the most common source of error in the plot command. The `plot` function also works with a single vector argument, in which case the elements of the vector are plotted against row or column indices. Thus, for two column vectors x and y each of length n,

`plot(x,y)`	plots y vs. x with a solid line (the default line style),
`plot(x,y,'--')`	plots y vs. x with a dashed line (more on this below),
`plot(x)`	plots the elements of x against their row index.

6.1.1   Style options

The *style-option* in the plot command is a character string that consists of 1, 2, or 3 characters that specify the color and/or the line style. There are several color, line-style, and marker-style options:

Color Style-option		*Line Style-option*		*Marker Style-option*	
y	yellow	–	solid	+	plus sign
m	magenta	--	dashed	o	circle
c	cyan	:	dotted	*	asterisk
r	red	-.	dash-dot	x	x-mark
g	green			.	point
b	blue			^	up triangle
w	white			s	square
k	black			d	diamond, etc.

The *style-option* is made up of either the color option, the line-style option, or a combination of the two.

Examples:

`plot(x,y,'r')`	plots y vs. x with a red solid line,
`plot(x,y,':')`	plots y vs. x with a dotted line,
`plot(x,y,'b--')`	plots y vs. x with a blue dashed line,
`plot(x,y,'+')`	plots y vs. x as unconnected points marked by +.

When no style option is specified, MATLAB uses the default option—a blue solid line.

6.1.2   Labels, title, legend, and other text objects

Plots may be annotated with `xlabel`, `ylabel`, `title`, and `text` commands.

The first three commands take string arguments, while the last one requires three arguments— `text(`*x-coordinate, y-coordinate, 'text'*`)`, where the coordinate values are taken from the current plot. Thus,

`xlabel('Pipe Length')`	labels the x-axis with `Pipe Length`,
`ylabel('Fluid Pressure')`	labels the y-axis with `Fluid Pressure`,

`title('Pressure Variation')`	titles the plot with `Pressure Variation`,
`text(2,6,'Note this dip')`	writes 'Note this dip' at the location (2.0,6.0) in the plot coordinates.

For on-line help type:
`help graph2d`

We have already seen an example of `xlabel`, `ylabel`, and `title` in Fig. 3.10. An example of `text` appears in Fig. 6.2. The arguments of `text`(*x,y, 'text'*) command may be vectors, in which case x and y must have the same length and *text* may be just one string or a vector of strings. If *text* is a vector then it must have the same length as x and, of course, like any other string vector, must have each element of the same length. A useful variant of the `text` command is `gtext`, which only takes string argument (a single string or a vector of strings) and lets the user specify the location of the text by clicking the mouse at the desired location in the graphics window.

Legend:

The `legend` command produces a boxed legend on a plot, as shown, for example, in Fig. 6.3 on page 166. The `legend` command is quite versatile. It can take several optional arguments. The most commonly used forms of the command are listed below.

`legend(`*string1, string2, ..*`)`	produces legend using the text in *string1, string2*, etc. as labels.
`legend(`*LineStyle1, string1, ..*`)`	specifies the line-style of each label.
`legend(.., `*pos*`)`	writes the legend outside the plot-frame if *pos* = -1 and inside the frame if *pos* = 0. (There are other options for *pos* too.)
`legend off`	deletes the legend from the plot.

When MATLAB is asked to produce a legend, it tries to find a place on the plot where it can write the specified legend without running into lines, grid, and other graphics objects. The optional argument *pos* specifies the location of the legend box. `pos=1` places the legend in the upper right hand corner (default), `2` in the upper left hand corner, `3` in the lower left hand corner, and `4` in the lower right hand corner. The user, however, can move the legend at will with the mouse (click and drag). For more information, see the on-line help on `legend`.

6.1.3   Axis control, zoom-in, and zoom-out

Once a plot is generated you can change the axes limits with the `axis` command. Typing

$$\boxed{\texttt{axis([}xmin\ xmax\ ymin\ ymax\texttt{])}}$$

changes the current axes limits to the specified new values $xmin$ and $xmax$ for the x-axis and $ymin$ and $ymax$ for the y-axis.

Examples:

`axis([-5 10 2 22]);` sets the x-axis from -5 to 10, y-axis from 2 to 22.
`axy = [-5 10 2 22]; axis(axy);`   same as above.
`ax = [-5 10]; ay = [2 22]; axis([ax ay]);`   same as above.

The `axis` command may thus be used to zoom-in on a particular section of the plot or to zoom-out[1]. There are also some useful predefined string arguments for the `axis` command:

`axis('equal')`   sets equal scale on both axes
`axis('square')`   sets the default rectangular frame to a square
`axis('normal')`   resets the axis to default values
`axis('axis')`   freezes the current axes limits
`axis('off')`   removes the surrounding frame and the tick marks.

The `axis` command must come after the `plot` command to have the desired effect.

Semi-control of axes

It is possible to control only part of the axes limits and let MATLAB set the other limits automatically. This is achieved by specifying the desired limits in the `axis` command along with `inf` as the values of the limits which you would like to be set automatically. For example,

`axis([-5 10 -inf inf])`   sets the x-axis limits at -5 and 10 and lets the y-axis limits be set automatically.
`axis([-5 inf -inf 22])`   sets the lower limit of the x-axis and the upper limit of the y-axis, and leaves the other two limits to be set automatically.

For on-line help type:
`help plotedit`
`help propedit`

6.1.4   Modifying plots with *Plot Editor*

MATLAB 6 provides an enhanced (over previous versions) interactive tool for modifying an existing plot. To activate this tool, go to the **Figure** window and click on the left-leaning *arrow* in the menu bar (see Fig. 6.1). Now you can select and double (or right) click on any object in the current plot to edit it. Double clicking on the selected object brings up a **Property Editor** window where you can select and modify the current properties of the object. Other tools in the menu bar, e.g., text (marked by **A**), arrow, and line, lets you modify and annotate figures just like simple graphics packages do.

---

[1]There is also a `zoom` command which can be used to zoom-in and zoom-out using the mouse in the figure window. See the on-line help on `zoom`.

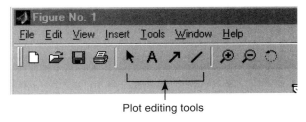

Plot editing tools

Figure 6.1: MATLAB provides interactive plot editing tools in the **Figure** window menu bar. Select the first arrow (left-leaning) for activating plot editor. Select **A**, the right-leaning arrow, and the diagonal line for adding text, arrows, and lines, respectively, in the current plot.

You can also activate the plot editor in the figure window by typing `plotedit` on the command prompt. You can activate the property editor by typing `propedit` at the command prompt. However, to make good use of the property editor, you must have some understanding of Handle Graphics. See Section 6.4 on page 190 for details.

6.1.5 Overlay plots

There are three different ways of generating overlay plots in MATLAB: the `plot`, `hold`, and `line` commands.

Method-1: Using the `plot` command to generate overlay plots

If the entire set of data is available, `plot` command with multiple arguments may be used to generate an overlay plot. For example, if we have three sets of data— (x1,y1), (x2,y2), and (x3,y3)—the command `plot(x1,y1, x2,y2,':', x3,y3,'o')` plots (x1,y1) with a solid line, (x2,y2) with a dotted line, and (x3,y3) as unconnected points marked by small circles ('o'), all on the same graph (See Fig. 6.2 for example). Note that the vectors (xi,yi) must have the same length pairwise. If the length of all vectors is the same, then it is convenient to make a matrix of **x** vectors and a matrix of **y** vectors and then use the two matrices as the argument of the `plot` command. For example, if x1, y1, x2, y2, x3, and y3 are all column vectors of length n then typing `X=[x1 x2 x3]; Y=[y1 y2 y3]; plot(X,Y)` produces a plot with three lines drawn in different colors. When `plot` command is used with matrix arguments, each column of the second argument matrix is plotted against the corresponding column of the first argument matrix.

```
>> t=linspace(0,2*pi,100);          % Generate vector t
>> y1=sin(t);   y2=t;               % Calculate y1, y2, y3
>> y3=t-(t.^3)/6+(t.^5)/120;
>> plot(t,y1,t,y2,'--',t,y3,'o')    % Plot (t,y1) with solid line
                                    %- (t,y2) with dahed line and
                                    %- (t,y3) with circles
>> axis([0 5 -1 5])                 % Zoom-in with new axis limits
>> xlabel('t')                      % Put x-label
>> ylabel('Approximations of sin(t)')% Put y-label
>> title('Fun with sin(t)')         % Put title
>> text(3.5,0,'sin(t)')             % Write 'sin(t)' at point (3.5,0)
>> gtext('Linear approximation')
>> gtext('First 3 terms')
>> gtext('in Taylor series')
```

gtext writes the specified string at a location clicked with the mouse in the graphics window. So after hitting return at the end of gtext command, go to the graphics window and click a location.

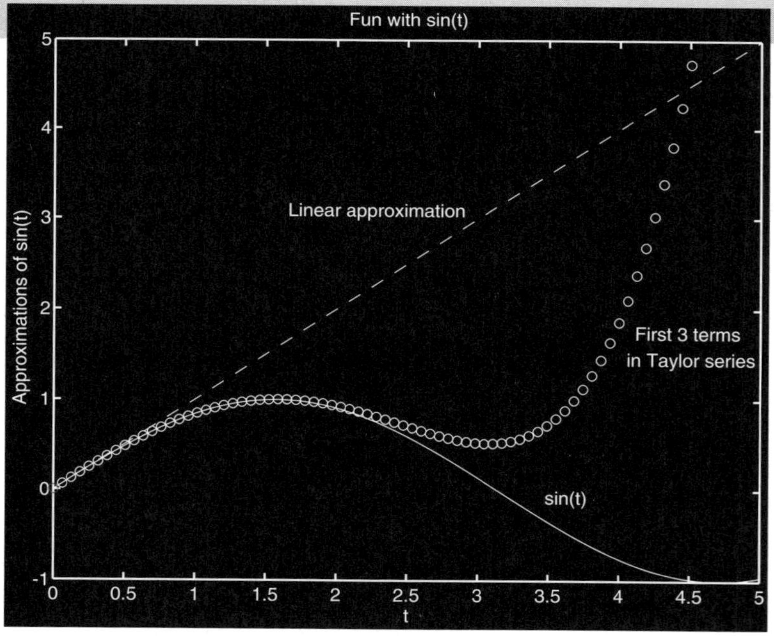

Figure 6.2: Example of an overlay plot along with examples of xlabel, ylabel, title, axis, text, and gtext commands. The three lines plotted are $y_1 = \sin t$, $y_2 = t$, and $y_3 = t - \frac{t^3}{3!} + \frac{t^5}{5!}$.

Method-2: Using the `hold` command to generate overlay plots

Another way of making overlay plots is with the `hold` command. Invoking `hold on` at any point during a session freezes the current plot in the graphics window. All subsequent plots generated by the `plot` command are simply added to the existing plot. The following script file shows how to generate the same plot as in Fig. 6.2 by using the `hold` command.

```
% - Script file to generate an overlay plot with the hold  command -
x=linspace(0,2*pi,100);       % Generate vector x
y1=sin(x);                    % Calculate y1
plot(x,y1)                    % Plot (x,y1) with solid line
hold on                       % Invoke hold for overlay plots
y2=x; plot(x,y2,'--')         % Plot (x,y2) with dashed line
y3=x-(x.^3)/6+(x.^5)/120;     % Calculate y3
plot(x,y3,'o')                % Plot (x,y3) as pts. marked by 'o'
axis([0 5 -1 5])              % Zoom-in with new axis limits
hold off                      % Clear hold command
```

The `hold` command is useful for overlay plots when the entire data set to be plotted is not available at the same time. You should use this command if you want to keep adding plots as the data becomes available. For example, if a set of calculations done in a `for` loop generates vectors x and y at the end of each loop and you would like to plot them on the same graph, `hold` is the way to do it.

Method-3: Using the `line` command to generate overlay plots

The `line` is a low-level graphics command which is used by the `plot` command to generate lines. Once a plot exists in the graphics window, additional lines may be added by using the `line` command directly. The `line` command takes a pair of vectors (or a triplet in 3-D) followed by *parameter name/parameter value* pairs as arguments:

```
line(xdata, ydata, ParameterName, ParameterValue)
```

This command simply adds lines to the existing axes. For example, the overlay plot created by the above script file could also be created with the following script file, which uses the `line` command instead of the `hold` command. As a bonus to the reader, we include an example of the `legend` command (see page 161).

```
% -- Script file to generate an overlay plot with the line command --
t=linspace(0,2*pi,100);       % Generate vector t
y1=sin(t);                    % Calculate y1, y2, y3
y2=t;
y3=t-(t.^3)/6+(t.^5)/120;
```

```
plot(t,y1)                    % Plot (t,y1) with (default) solid line
line(t,y2,'linestyle','--')   % Add line (t,y2) with dahed line and
line(t,y3,'marker','o')       % Add line (t,y3) plotted with circles

axis([0 5 -1 5])              % Zoom-in with new axis limits

xlabel('t')                   % Put x-label
ylabel('Approximations of sin(t)')
                              % Put y-label
title('Fun with sin(t)')      % Put title

legend('sin(t)','linear approx.','5th order approx.')
                              % add legend
```

The output generated by the above script file is shown in Fig. 6.3. After generating the plot, click and hold the mouse on the legend rectangle and see if you can drag the legend to some other position. Alternatively, you could specify an option in the legend command to place the legend rectangle in any of the four corners of the plot. See the on-line help on legend.

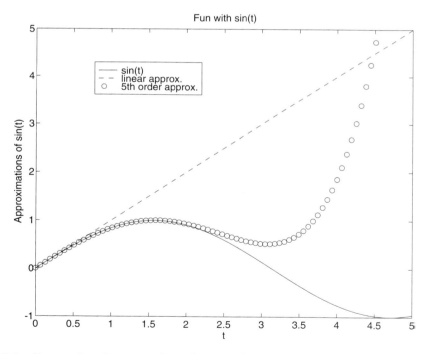

Figure 6.3: Example of an overlay plot produced by using the line command. The legend is produced by the legend command. See the script file for details.

6.1.6 Specialized 2-D plots

There are many specialized graphics functions for 2-D plotting. They are used as alternatives to the `plot` command we have just discussed. There is a whole suite of *ezplotter* functions, such as `ezplot`, `ezpolar`, `ezcontour`, etc., which are truly easy to use. See Section 3.6 for a discussion and examples of these functions.

Here, we provide a list of other functions commonly used for plotting *x-y* data:

> *For on-line help type:*
> `help graph2d,`
> `help specgraph`

`area`	creates a filled area plot
`bar`	creates a bar graph
`barh`	creates a horizontal bar graph
`comet`	makes an animated 2-D plot
`compass`	creates arrow graph for complex numbers
`contour`	makes contour plots
`contourf`	makes filled contour plots
`errorbar`	plots a graph and puts error bars
`feather`	makes a feather plot
`fill`	draws filled polygons of specified color
`fplot`	plots a function of a single variable
`hist`	makes histograms
`loglog`	creates plot with log scale on both x and y axes
`pareto`	makes pareto plots
`pcolor`	makes pseudocolor plot of a matrix
`pie`	creates a pie chart
`plotyy`	makes a double y-axis plot
`plotmatrix`	makes a scatter plot of a matrix
`polar`	plots curves in polar coordinates
`quiver`	plots vector fields
`rose`	makes angled histograms
`scatter`	creates a scatter plot
`semilogx`	makes semilog plot with log scale on the x-axis
`semilogy`	makes semilog plot with log scale on the y-axis
`stairs`	plots a stair graph
`stem`	plots a stem graph

On the following pages we show examples of these functions. The commands shown in the middle column produce the plots shown in the right column. There are several ways you can use these graphics functions. Also, many of them take optional arguments. The following examples should give you a basic idea of how to use these functions and what kind of plot to expect from them. For more information on any of these functions see the on-line help.

Function	Example Script	Output
fplot	$f(t) = t\sin t, \;\; 0 \le t \le 10\pi$ `fplot('x.*sin(x)',[0 10*pi])` Note that the function to be plotted must be written as a function of x.	

semilogx	$x = e^{-t}, \;\; y = t, \; 0 \le t \le 2\pi$
	
	 `t=linspace(0,2*pi,200);`
	 `x = exp(-t); y = t;`
	 `semilogx(x,y), grid`	

semilogy	$x = t, \; y = e^{t}, \;\; 0 \le t \le 2\pi$
	
	 `t=linspace(0,2*pi,200);`
	 `semilogy(t,exp(t))`
	 `grid`	

loglog	$x = e^{t}, \; y = 100 + e^{2t}, \;\; 0 \le t \le 2\pi$
	
	 `t=linspace(0,2*pi,200);`
	 `x = exp(t);`
	 `y = 100+exp(2*t);`
	 `loglog(x,y), grid`	

Function	Example Script	Output

polar	$r^2 = 2\sin 5t, \;\; 0 \le t \le 2\pi$
	
	```t=linspace(0,2*pi,200);```
	```r=sqrt(abs(2*sin(5*t)));```
	```polar(t,r)```	

fill	$\begin{aligned} r^2 &= 2\sin 5t, \;\; 0 \le t \le 2\pi \\ x &= r\cos t, \;\; y = r\sin t \end{aligned}$
	
	```t=linspace(0,2*pi,200);```
	```r=sqrt(abs(2*sin(5*t)));```
	```x=r.*cos(t);```
	```y=r.*sin(t);```
	```fill(x,y,'k'),```
	```axis('square')```	

bar	$\begin{aligned} r^2 &= 2\sin 5t, \;\; 0 \le t \le 2\pi \\ y &= r\sin t \end{aligned}$
	
	```t=linspace(0,2*pi,200);```
	```r=sqrt(abs(2*sin(5*t)));```
	```y=r.*sin(t);```
	```bar(t,y)```
	```axis([0 pi 0 inf]);```	

errorbar	$\begin{aligned} f_{\text{approx}} &= x - \frac{x^3}{3!}, \;\; 0 \le x \le 2 \\ error &= f_{\text{approx}} - \sin x \end{aligned}$
	
	```x=0:.1:2;```
	```aprx2=x-x.^3/6;```
	```er=aprx2-sin(x);```
	```errorbar(x,aprx2,er)```	

Function	Example Script	Output

hist	World population by continents
	
	``matlab
	cont=char('Asia','Europe','Africa',...
	     'N. America','S. America');
	pop = [3332;696;694;437;307];
	barh(pop)
	for i=1:5,
	     gtext(cont(i,:));
	end
	xlabel('Population in millions')
	Title('World Population (1992)',...
	     'fontsize',18)
	``	

plotyy	$$y_1 = e^{-x}\sin x,\ 0 \le t \le 10$$
	$$y_2 = e^{x}$$
	
	``matlab
	x=1:.1:10;
	y1 = exp(-x).*sin(x);
	y2 = exp(x);
	Ax = plotyy(x,y1,x,y2);
	hy1 = get(Ax(1),'ylabel');
	hy2 = get(Ax(2),'ylabel');
	set(hy1,'string','e^-x sin(x)');
	set(hy2,'string','e^x ');
	``	

area	$$y = \frac{\sin(x)}{x},\quad -3\pi \le x \le 3\pi$$
	
	``matlab
	x=linspace(-3*pi,3*pi,100);
	y=-sin(x)./x;
	area(x,y)
	xlabel('x'),ylabel('sin(x)./x')
	hold on
	x1=x(46:55); y1=y(46:55);
	area(x1,y1,'facecolor','y')
	``	

pie	World population by continents
	
	``matlab
	cont=char('Asia','Europe','Africa',...
	     'N. America','S. America');
	pop = [3332;696;694;437;307];
	pie(pop)
	for i=1:5,
	     gtext(cont(i,:));
	end
	Title('World Population',...
	     'fontsize',18)
	``	

Function	Example Script	Output

hist	Histogram of 50 randomly distributed numbers between 0 and 1.
	
	`y=randn(50,1);`
	`hist(y)`	

stem	$f = e^{-t/5} \sin t,\ 0 \le t \le 2\pi$
	
	`t=linspace(0,2*pi,200);`
	`f=exp(-.2*t).*sin(t);`
	`stem(t,f)`	

stairs	$\begin{aligned} r^2 &= 2\sin 5t,\ \ 0 \le t \le 2\pi \\ y &= r\sin t \end{aligned}$
	
	`t=linspace(0,2*pi,200);`
	`r=sqrt(abs(2*sin(5*t)));`
	`y=r.*sin(t);`
	`stairs(t,y)`
	`axis([0 pi 0 inf]);`	

compass	$z = \cos\theta + i\sin\theta,\ \ -\pi \le \theta \le \pi$
	
	`th=-pi:pi/5:pi;`
	`zx=cos(th);`
	`zy=sin(th);`
	`z=zx+i*zy;`
	`compass(z)`	

Function	Example Script	Output

comet	$y = t\sin t,\ \ 0 \leq t \leq 10\pi$
	
	`q=linspace(0,10*pi,200);`
	`y = q.*sin(q);`
	`comet(q,y)`
	(Its better to see it on screen)	

contour	$z \ = \ -\dfrac{1}{2}x^2 + xy + y^2$
	$\|x\| \leq 5,\ \|y\| \leq 5.$
	
	`r=-5:.2:5;`
	`[X,Y]=meshgrid(r,r);`
	`Z=-.5*X.^2 + X.*Y + Y.^2;`
	`cs=contour(X,Y,Z);`
	`clabel(cs)`	

quiver	$z \ = \ x^2 + y^2 - 5\sin(xy)$
	$\|x\| \leq 2,\ \|y\| \leq 2.$
	
	`r=-2:.2:2;`
	`[X,Y]=meshgrid(r,r);`
	`Z=X.^2 -5*sin(X.*Y) + Y.^2;`
	`[dx,dy]=gradient(Z,.2,.2);`
	`quiver(X,Y,dx,dy,2);`	

pcolor	$z \ = \ x^2 + y^2 - 5\sin(xy)$
	$\|x\| \leq 2,\ \|y\| \leq 2.$
	
	`r=-2:.2:2;`
	`[X,Y]=meshgrid(r,r);`
	`Z=X.^2 -5*sin(X.*Y) + Y.^2;`
	`pcolor(Z), axis('off')`
	`shading interp`	

6.2 Using `subplot` to Layout Multiple Graphs

If you want to make a few plots and place the plots side-by-side (not overlay), use the `subplot` command to design your layout. The subplot command requires three integer arguments:

$$\boxed{\texttt{subplot(m,n,p)}}$$

Subplot divides the graphics window into $m \times n$ sub-windows and puts the plot generated by the next plotting command into the pth sub-window where the sub-windows are counted row-wise. Thus, the command `subplot(2,2,3)`, `plot(x,y)` divides the graphics window into 4 sub-windows and plots y vs. x in the 3rd sub-window, which is the first sub-window in the second row. See Fig. 6.5.

6.3 3-D Plots

For on-line help type:
`help graph3d`

MATLAB provides extensive facilities for visualization of 3-D data. In fact, the built-in *colormaps* may be used to represent the 4th dimension. The facilities provided include built-in functions for plotting space-curves, wire-frame objects, surfaces and shaded surfaces, generating contours automatically, displaying volumetric data, specifying light sources, interpolating colors and shading, and even displaying images. Typing `help graph3d` in the command window gives a list of functions available for general 3-D graphics. Here is a list of commonly used functions other than those from the **ez**-stable, such as `ezsurf`, `ezmesh`, `ezplot3`, etc., discussed in Section 3.6.

`plot3`	plots curves in space
`stem3`	creates discrete data plot with stems in 3-D
`bar3`	plots 3-D bar graph
`bar3h`	plots 3-D horizontal bar graph
`pie3`	makes 3-D pie chart
`comet3`	makes animated 3-D line plot
`fill3`	draws filled 3-D polygons
`contour3`	makes 3-D contour plots
`quiver3`	draws vector fields in 3D
`scatter3`	makes scatter plot in 3-D
`mesh`	draws 3-D mesh surfaces (wire-frame)
`meshc`	draws 3-D mesh surface along with contours
`meshz`	draws 3-D mesh surface with curtain plot of reference planes
`surf`	creates 3-D surface plots
`surfc`	creates 3-D surface plots along with contours
`surfl`	creates 3-D surface plots with specified light source
`trimesh`	mesh plot with triangles

trisurf	surface plot with triangles
slice	draws a volumetric surface with slices
waterfall	creates a *waterfall* plot of 3-D data
cylinder	generates a cylinder
ellipsoid	generates an ellipsoid
sphere	generates a sphere

Among these functions, plot3 and comet3 are the 3-D analogues of plot and comet commands mentioned in the 2-D graphics section. The general syntax for the plot3 command is

$$\boxed{\texttt{plot3}(x,\ y,\ z,\ \texttt{'style-option'})}$$

This command plots a curve in 3-D space with the specified line style. The argument list can be repeated to make overlay plots, just the same way as with the plot command. A catalog of these functions with example scripts and the corresponding output is given on pages 181–185. Since the example scripts use a few functions which we have not discussed yet, we postpone the catalog till we discuss these functions.

Plots in 3-D may be annotated with functions already mentioned for 2-D plots— xlabel, ylabel, title, text, grid, etc., along with the obvious addition of zlabel. The grid command in 3-D makes the 3-D appearance of the plots better, especially for curves in space (see Fig. 6.5 for example).

6.3.1 View

The viewing angle of the observer is specified by the command

$$\boxed{\texttt{view}(azimuth,\ elevation)}$$

where *azimuth*, in degrees, specifies the horizontal rotation from the y-axis, measured positive counterclockwise, and *elevation*, in degrees, specifies the vertical angle measured positive above the xy-plane (see Fig. 6.4). The default values for these angles are $-37.5°$ and $30°$ respectively.

By specifying appropriate values of the azimuth and the elevation, one can plot the projections of a three-dimensional object on different 2-D planes. For example, the command view(90,0) puts the viewer on the positive x-axis, looking straight on the yz-plane, and thus produces a 2-D projection of the object on the yz-plane. Figure 6.5 shows the projections obtained by specifying view angles. The script file that generates the data, plots the curves, and obtains the different projected views is listed below.

```
%--- Script file with examples of 'subplot' and 'view' commands -—
clf                    % Clear any previous figure
t=linspace(0,6*pi,100); % Generate vector t with 100
                       %- equally spaced points between 0 and 6*pi
x=cos(t); y=sin(t); z=t; % Calculate x, y, and z
```

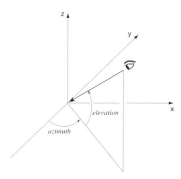

Figure 6.4: The viewing angles *azimuth* and *elevation* in 3-D plots.

```
subplot(2,2,1)              % Divide the graphics window into 4
                           %- subwindows and plot in the first window
plot3(x,y,z),grid          % Plot the space curve in 3-D and put grid
xlabel('cos(t)'),ylabel('sin(t)'),zlabel('t')
title('A circular helix: x(t)=cos(t), y(t)=sin(t), z(t)=t')

subplot(2,2,2)              % get figure subwindow 2 ready
plot3(x,y,z), view(0,90),% View along the z-axis from above
xlabel('cos(t)'),ylabel('sin(t)'),zlabel('t')
title('Projection in the X-Y plane')

subplot(2,2,3)              % get figure subwindow 3 ready
plot3(x,y,z), view(0,0), % View along the y-axis
xlabel('cos(t)'),ylabel('sin(t)'),zlabel('t')
title('Projection in the X-Z plane')

subplot(2,2,4)              % get figure subwindow 4 ready
plot3(x,y,z),view(90,0),% View along the x-axis
xlabel('cos(t)'),ylabel('sin(t)'),zlabel('t')
title('Projection in the Y-Z plane')
```

View(2) and View(3)

These are the special cases of the **view** command, specifying the default 2-D and 3-D views:

view(2)   same as view(0,90), shows the projection in the xz-plane.
view(3)   same as view(-37.5,30), shows the default 3-D view.

The **view(3)** command can be used to see a 2-D object in 3-D. It may be useful in visualizing the perspectives of different geometrical shapes. The following script file draws a filled circle in 2-D and also views the same circle in 3-D. The output is shown in Fig. 6.6

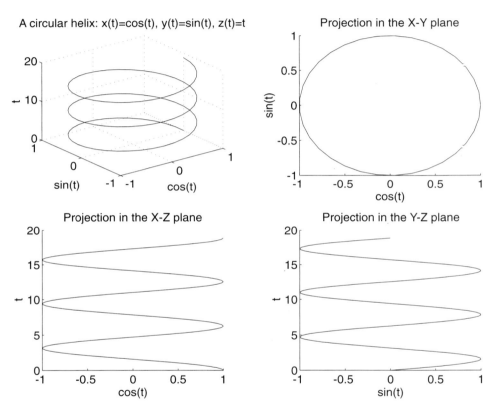

Figure 6.5: Examples of `plot3`, `subplot` and `view`. Note how the 3-D grid in the background helps in the 3-D appearance of the space curve. Although the three 2-D pictures could be made using the `plot` command, this example illustrates the use of viewing angles.

```
% ----- script file to draw a filled circle and view in 3D -----
theta = linspace(0,2*pi,100);     % create vector theta
x = cos(theta);                   % generate x-coordinates
y = sin(theta);                   % generate y-coordinates
subplot(1,2,1)                    % initiate a 1 by 2 subplot
fill(x,y,'g');axis('square');     % plot the filled circle
subplot(1,2,2)                    % go to the 2nd subplot
fill(x,y,'g');axis('square');     % plot the same circle again
view(3)                           % view the 2-D circle in 3-D
```

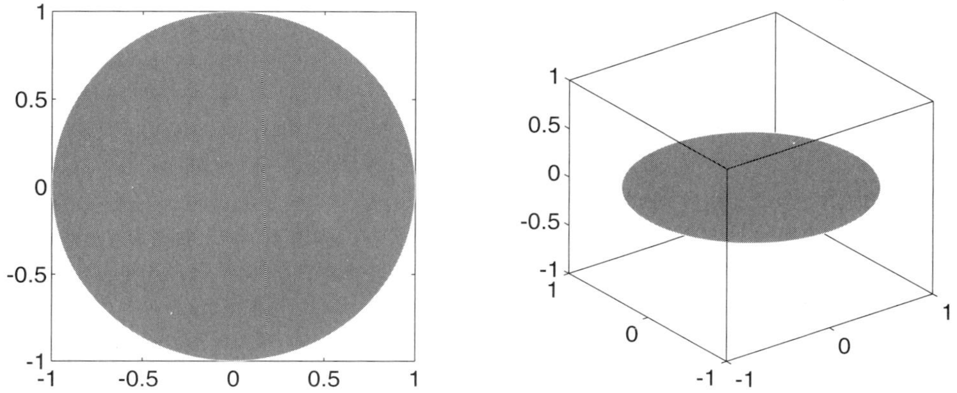

Figure 6.6: Example of `view(3)` to see a 2-D object in 3-D.

6.3.2 Rotate view

MATLAB 5 provides more versatile and easier utilities for manipulating the view angle of a 3-D plot. There is a simple utility called **rotate3d** . Simply turn it on with **rotate3d on** and rotate the view with your mouse. There are also some sophisticated *camera* functions that let you specify the camera view angle, zoom, roll, pan, etc. See the on-line help on **graph3D**.

6.3.3 Mesh and surface plots

The functions for plotting meshes and surfaces **mesh** and **surf**, and their various variants **meshz, meshc, surfc**, and **surfl**, take multiple optional arguments, the most basic form being **mesh(Z)** or **surf(Z)**, where Z represents a matrix. Usually surfaces are represented by the values of z-coordinates sampled on a grid of (x, y) values. Therefore, to create a surface plot we first need to generate a grid of (x, y) coordinates and find the height (z-coordinate) of the surface at each of the grid points. Note that you need to do the same thing for plotting a function of two variables. MATLAB provides a function **meshgrid** to create a grid of points over a specified range.

The function meshgrid: Suppose we want to plot a function $z = x^2 - y^2$ over the domain $0 \leq x \leq 4$ and $-4 \leq y \leq 4$. To do so, we first take several points in the domain, say 25 points, as shown in Fig. 6.7. We can create two matrices X and Y, each of size 5×5, and write the xy-coordinates of each point in these matrices. We can then evaluate z with the command z = X.^2-Y.^2;. Creating the two matrices X and Y is much easier with the meshgrid command:

```
rx = 0:4;               % create a vector rx=[0 1 2 3 4]
ry = -4:2:4;            % create a vector ry=[-4 -2 0 2 4]
[X,Y] = meshgrid(rx,ry); % create a grid of 25 points and
                        %- store their coordinates in X and Y.
```

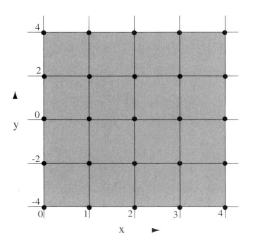

Figure 6.7: A grid of 25 points in the xy-plane. The grid can be created by the meshgrid command: [X,Y] = meshgrid(rx,ry); where rx and ry are vectors specifying the location of grid lines along x and y axes.

The commands shown above generate the 25 points shown in Fig. 6.7. All we need to generate is two vectors, rx and ry, to define the region of interest and distribution of grid points. Also, the two vectors need not be either same sized or linearly spaced (although, most of the time, we take square regions and create grid points equally spaced in both directions. See examples on pages 181–185). To be comfortable with 3-D graphics, you should understand the use of meshgrid.

Back to mesh **plot:** When a surface is plotted with mesh(z) (or surf(z)) command, where z is a matrix, then the tickmarks on the x and y axes do not indicate the domain of z but the row and column indices of the z-matrix. This is the default. Typing mesh(x,y,z) or surf(x,y,z), where x and y are vectors used by meshgrid command to create a grid, results in the surface plot of z with x- and y-values shown on the x and y axes. The following script file should serve as an example of how to use `meshgrid` and `mesh` command. Here we try to plot the surface

$$z = \frac{xy(x^2 - y^2)}{x^2 + y^2}, \quad -3 \le x \le 3, \ -3 \le y \le 3$$

by computing the values of z over a 50×50 grid on the specified domain. The results of the two plot commands are shown in Fig. 6.8.

```
%-------------------------------------------------------------
% Script file to generate and plot the surface
% z =  xy(x^2-y^2)/(x^2+y^2) using meshgrid and mesh commands.
%-------------------------------------------------------------
x=linspace(-3,3,50); y=x;  % Generate 50 element long vectors x & y
[X,Y]=meshgrid(x,y);       % Create a grid over the specified domain
Z=X.*Y.*(X.^2-Y.^2)./(X.^2+Y.^2);   % Calculate Z at each grid point
mesh(X,Y,Z)                % Make a wire-frame surface plot of Z and
                           %- use x and y values on the x and y-axes.
figure(2)                  % Open a new figure window
meshc(X,Y,Z),view(-55,20)  % Plot the same surface along with
                           %- contours and show the view from
                           %- the specified angles.
```

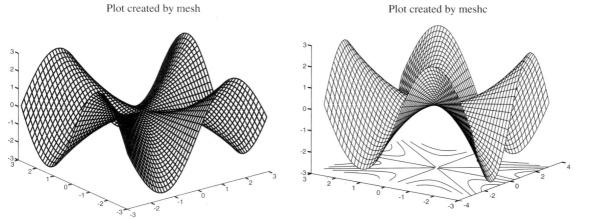

Figure 6.8: 3-D surface plots created by `mesh` and `meshc` commands. The second plot uses a different viewing angle to show the center of the contour lines. Note that the surfaces do not show hidden lines (this is the default setting; it can be changed with the `hidden` command).

While surfaces created by `mesh` or its variants have a wire-frame appearance, surfaces created by the `surf` command or its variants produce a true surface-like appearance, especially when used with the `shading` command.

There are three kinds of shading available— `shading flat` produces simple flat shading, `shading interp` produces more dramatic interpolated shading, and `shading faceted`, the default shading, shows shaded facets of the surface. Both `mesh` and `surf` can plot parametric surfaces with color scaling to indicate a fourth dimension. This is accomplished by giving four matrix arguments to these commands, e.g. `surf(X,Y,Z,C)` where X, Y, and Z are matrices representing a surface in parametric form and C is the matrix indicating color scaling. The command `surfl` can be used to control the light reflectance and to produce special effects with a specified location of a light source. See on line help on `surfl` for more information.

We close this section with a catalog of the popular 3D graphics functions. We hope that you can use these functions for your needs simply by following the example scripts. But we acknowledge that the `meshgrid` command takes some thought to understand well.

Function	Example Script	Output
plot3	Plot of a parametric space curve: $$x(t) = t, \ y(t) = t^2, \ z(t) = t^3.$$ $$0 \le t \le 1.$$ `t=linspace(0,1,100);` `x=t; y=t.^2; z=t.^3;` `plot3(x,y,z),grid`	
fill3	Plot of 4 filled polygons with 3 vertices each. `X=[0 0 0 0; 1 1 -1 1;` `   1 -1 -1 -1];` `Y=[0 0 0 0; 4 4 4 4;` `   4 4 4 4];` `Z=[0 0 0 0; 1 1 -1 -1;` `   -1 1 1 -1];` `fill3(X,Y,Z,rand(3,4))` `view(120,30)`	
contour3	Plot of 3-D contour lines of $$z = -\frac{5}{1 + x^2 + y^2},$$ $$|x| \le 3, |y| \le 3.$$ `r = linspace(-3,3,50);` `[x,y]=meshgrid(r,r);` `z=-5./(1+x.^2+y.^2);` `contour3(z)`	

Function	Example Script	Output

surf	$z = \cos x \, \cos y \, e^{\frac{-\sqrt{x^2+y^2}}{4}}$
	 $\|x\| \le 5, \quad \|y\| \le 5$
	
	 ```u = -5:.2:5;```
	```[X,Y] = meshgrid(u, u);```
	```Z = cos(X).*cos(Y).*...```
	``` exp(-sqrt(X.^2+Y.^2)/4);```
	```surf(X,Y,Z)```	

surfc	$z = \cos x \, \cos y \, e^{\frac{-\sqrt{x^2+y^2}}{4}}$
	 $\|x\| \le 5, \quad \|y\| \le 5$
	
	 ```u = -5:.2:5;```
	```[X,Y] = meshgrid(u, u);```
	```Z = cos(X).*cos(Y).*...```
	``` exp(-sqrt(X.^2+Y.^2)/4);```
	```surfc(Z)```
	```view(-37.5,20)```
	```axis('off')```	

surfl	$z = \cos x \, \cos y \, e^{\frac{-\sqrt{x^2+y^2}}{4}}$
	 $\|x\| \le 5, \quad \|y\| \le 5$
	
	 ```u = -5:.2:5;```
	```[X,Y] = meshgrid(u, u);```
	```Z = cos(X).*cos(Y).*...```
	``` exp(-sqrt(X.^2+Y.^2)/4);```
	```surfl(Z)```
	```shading interp```
	```colormap hot```	

Note: Plotting a surface with `surf(X,Y,Z)` shows proper values on the x and y axes while plotting the surface with `surf(Z)` shows the row and column indices of matrix `Z` on the x and y axes. Same is true for other 3D plotting commands such as `mesh`, `contour3`, etc. Compare the values on the x and y axes in the first and the last figure in the table above.

Function	Example Script	Output

mesh	$$z = -\frac{5}{1 + x^2 + y^2}$$ $$|x| \le 3, \quad |y| \le 3$$
	
	 ```x = linspace(-3,3,50);```
	```y = x;```
	```[x,y] = meshgrid(x,y);```
	```z=-5./(1+x.^2+y.^2);```
	```mesh(z)```	

meshz	$$z = -\frac{5}{1 + x^2 + y^2}$$ $$|x| \le 3, \quad |y| \le 3$$
	
	 ```x = linspace(-3,3,50);```
	```y = x;```
	```[x,y] = meshgrid(x,y);```
	```z=-5./(1+x.^2+y.^2);```
	```meshz(z)```
	```view(-37.5, 50)```	

waterfall	$$z = -\frac{5}{1 + x^2 + y^2}$$ $$|x| \le 3, \quad |y| \le 3$$
	
	 ```x = linspace(-3,3,50);```
	```y = x;```
	```[x,y] = meshgrid(x,y);```
	```z=-5./(1+x.^2+y.^2);```
	```waterfall(z)```
	```hidden off```	

Function	Example Script	Output

pie3	World population by continents
	
	`% popdata:  As,Eu,Af,NA,SA`
	`pop=[3332;696;694;437;307];`
	`pie3(pop)`
	`Title('World Population')`	

stem3	Discrete data plot with stems
	
	$$x = t, \quad y = t\sin(t),$$
	$$z = e^{t/10} - 1$$
	for $0 \le t \le 6\pi$.
	
	`t=linspace(0,6*pi,200);`
	`x=t; y=t.*sin(t);`
	`z=exp(t/10)-1;`
	`stem3(x,y,z,'filled')`
	`xlabel('x'),`
	`ylabel('x sin(x)')`
	`zlabel('e^t/10-1')`	

ribbon	2D curves as ribbons in 3D
	
	$$y_1 = \sin(t), \quad y_2 = e^{.15t}\sin(t)$$
	$$y_3 = e^{.8t}\sin(t)$$
	for $0 \le t \le 5\pi$.
	
	`t=linspace(0,5*pi,100);`
	`y1 = sin(t);`
	`y2 = exp(-.15*t).*sin(t);`
	`y3 = exp(-.8*t).*sin(t);`
	`y = [y1; y2; y3];`
	`ribbon(t',y',.1)`	

Function	Example Script	Output
sphere	A unit sphere centered at the origin and generated by 3 matrices x, y, and z of size 21×21 each. ```\nsphere(20)\nor\n[x,y,z]=sphere(20);\nsurf(x,y,z)\n```	
ellipsoid	An ellipsoid of radii $rx = 1$, $ry = 2$ and $rz = 0.5$, centered at the origin. ```\ncx=0; cy=0; cz=0;\nrx=1; ry=2; rz=0.5;\nellipsoid(cx,cy,cz,rx,ry,rz)\n```	
cylinder	A cylinder generated by $$r = \sin(3\pi z) + 2$$ $$0 \le z \le 1, \quad 0 \le \theta \le 2\pi.$$ ```\nz=[0:.02:1]';\nr=sin(3*pi*z)+2;\ncylinder(r)\n```	
slice	Slices of the volumetric function $f(x,y,z) = x^2 + y^2 - z^2$ $|x| \le 3$, $|y| \le 3$, $|z| \le 3$ at $x = -2$ and 2.5, $y = 2.5$, and $z = 0$. ```\nv = [-3:.2:3];\n[x,y,z]=meshgrid(v,v,v);\nf=(x.^2+y.^2-z.^2);\nxv=[-2 2.5]; yv=2.5; zv=0;\nslice(x,y,z,f,xv,yv,zv);\nview([-30 30])\n``` The value of the function is indicated by the color intensity.	

For on-line help type:
`help vissuite`

6.3.4 Vector field and volumetric plots

One of the most crucial needs of visualization in scientific computation is for data that is essentially volumetric, i.e., defined over a three dimensional space. For example, we may have temperature or pressure defined over each (x, y, z) triple in a bounded 3-D space. How do we display this data graphically? If we have a function $z = f(x, y)$ defined over some finite region of the xy-plane, we can display z or f as a 3-D surface. But, we have $f(x, y, z)$! So, we need a 4-D hypersurface. That is the basic problem.

We display volumetric data by slicing it along several planes in 3-D and plotting the data on those planes either with graded color maps or with contours. Such displays are still an area of active research. However, we can do fairly well with the tools currently available. One of the most common applications is in the area of 3-D vector fields. A vector field defines a vector quantity as a function of the space variables (x, y, z). Fortunately, in this case we can display the data (the vector) with an arrow drawn at each (x, y, z) triple, with the magnitude of the vector represented by the length of the arrow, and the direction represented by the orientation of the arrow. This concept is used extensively in dynamical systems in various ways.

MATLAB 6 provides extensive tools for visualizing vector fields and volumetric data. Unfortunately, these tools are beyond the scope of this book. Therefore, we merely mention the tools here and give examples of only the 'most likely to be used' tools.

Plotting vector fields

The functions available for vector field visualization include `quiver`, `stream2`, `stream3`, `streamline`, `quiver3`, `coneplot`, `divergence`, `curl`, `streamtube`, `streamribbon`, `streamslice`, `streamparticles`, etc.

If $u(x, y)$ and $v(x, y)$ are given as vector components in x and y directions, respectively, then the vector field can be easily drawn with `quiver` (`quiver3` in 3-D). An example of `quiver` appears on page 167 in the table of 2-D plots. The *stream* functions are an extension of the same concept; they draw streamlines or trajectories from user specified points in the specified vector field. These suite of functions is a welcome addition in MATLAB 6.

The function `streamline` is useful for drawing solution trajectories in 2-D and 3-D vector fields defined by ODEs! You need not solve the ODEs! *Example:* Let (see also Exercise 8 on page 157)

$$\begin{aligned} \dot{x} &= y + x - x(x^2 + y^2) \\ \dot{y} &= -x + y - y(x^2 + y^2). \end{aligned}$$

These two ODEs define a vector field ($u \equiv \dot{x}$ and $v \equiv \dot{y}$). Let us use streamline to draw a few solution trajectories starting from various points in the xy-plane (initial conditions in the phase plane). The general syntax of streamline is

```
streamline(x,y,z, u,v,w, x0,y0,z0)
```

where (x,y,z) are 3-D matrices of grid-points where the vector field components (u,v,w) are specified, and (x0,y0,z0) are starting points for the streamlines to be drawn. Here is a script file that draws the streamlines for our 2-D vector field.

```
%STREAMLINE2D example of using streamline for 2-D vector field
% The vector field is given by two ODEs
% -----------------------------------
  %create grid points in 2-D
    v = linspace(-2,2,50);
    [X,Y] = meshgrid(x,y);
  %define vector field
    U = Y + X - X.*(X.^2 + Y.^2);
    V = Y - X - X.*(X.^2 + Y.^2);
  %specify starting points for streamlines
    x0 = [-2 -2 -2 -2 -.5 -.5 .5 .5 2 2 2 2 -.1 -.1 .1 .1];
    y0 = [-2 -.5 .5 2 -2 2 -2 2 -2 -.5 .5 2 -.1 .1 -.1 .1];
  %draw streamlines
    streamline(X,Y,U,V,X0,Y0)
```

The result obtained is shown in Fig. 6.9.

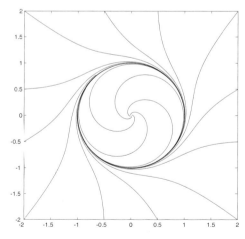

Figure 6.9: Plot obtained by executing the script file streamline2D

Plotting volumetric data

The functions available for volumetric data visualization include slice, slicecontour, isosurface, isonormal, isocaps, isocolors, subvolume, reducevolume, smooth3, reducepath, etc. See on-line help for more details.

6.3.5 Interpolated surface plots

Many a times, we get data (usually from experiments) in the form of (x, y, z) triples, and we want to fit a surface through the data. Thus, we have a vector z that contains the z-values corresponding to irregularly spaced x and y values. Here, we do not have a regular grid, as created by `meshgrid`, and we do not have a matrix Z that contains the z-values of the surface at those grid points. Therefore, we have to fit a surface through the given triplets (x_i, y_i, z_i). The task is much simpler than it seems. MATLAB 5 provides a function, `griddata`, that does this interpolation for us. The general syntax of this function is

$$[\texttt{Xi,Yi,Zi}] = \texttt{griddata(x,y,z,xi,yi,}\textit{method}\texttt{)}$$

where `x`, `y`, `z` are the given data vectors (non-uniformly spaced), `xi` and `yi` are the user prescribed points (hopefully, uniformly spaced) at which `zi` are computed by interpolation, and *method* is the choice for the interpolation algorithms. The algorithms available are *nearest*, *linear*, *cubic*, and *v4*. See the on-line documentation for description of these methods.

As an example let us consider 50 randomly distributed points in the xy-plane, in the range $-1 < x < 1$, and $-1 < y < 1$. Let the z values at these points be given by $z = 3/(1+x^2+y^2)$. Thus, we have three vectors of length 50 each. The data points are shown in Fig. 6.10 by the stem plot. Now, let us fit a surface through these points:

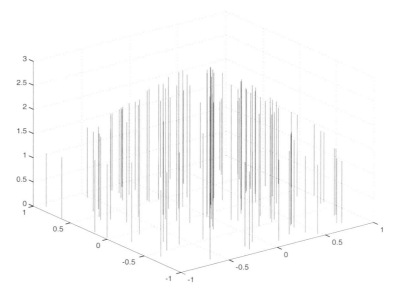

Figure 6.10: Non-uniformly distributed data points (x, y, z).

```
% SURFINTERP: script file to generate an interpolated surface
```

```
% Given vectors x, y, and z, generate data matrix Zi from
% interpolation to fit a surface through the data
% -----------------------------------------------------
xv=2*rand(1,100)-1;      % this is the given x
yv=2*rand(1,100)-1;      % this is the given y
zv=3./(1+xv.^2+yv.^2);   % this is the given z
stem3(xv,yv,zv)          % show data as stem plot

xi=linspace(-1,1,30);    % create uniformly spaced xi
yi=xi';                  % create uniformly spaced yi
                         % note that yi is a column

[Xi,Yi,Zi] = griddata(xv,yv,zv,xi,yi,'v4');
                         % interpolate surface using
                         % v4 (MATLAB 4 griddata) method
surf(Xi,Yi,Zi)           % plot the interpolated surface
```

The interpolated surface is shown in Fig. 6.11.

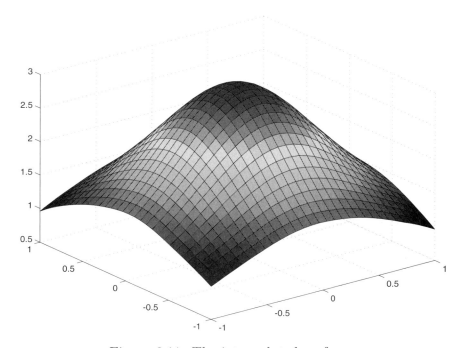

Figure 6.11: The interpolated surface.

For on-line help
type:
`help graphics`

6.4   Handle Graphics

> You need not learn or understand Handle Graphics to do most of the plotting an average person needs. If you want extra-detailed control of your graph appearance or want to do animation (beyond `comet` plots) you might want to learn Handle Graphics. This is NOT a topic for beginners.

A line is a graphics object. It has several properties—line style, color, thickness, visibility, etc. Once a line is drawn on the graphics screen, it is possible to change any of its properties later. Suppose you draw several lines with pencil on a paper. If you want to change one of the lines, you first must find the line you want to change and then change what you do not like about it. On the graphics screen, a line may be one among several graphics objects (e.g., axes, text, labels, etc.). So how do you get hold of a line? You get hold of a line by its *handle*.

What is a handle? MATLAB assigns a floating-point number to every object in the figure window (including invisible objects), and it uses this number as an address or name for the object in the figure. This number is the handle of the object.

Once you get hold of the handle, you can access all properties of the object. In short, the handle identifies the object and the object brings with it the list of its properties. In programming, this approach of defining objects and their properties is called *object-oriented programming*. The advantage it offers is that you can access individual objects and their properties and change any property of an object without affecting other properties or objects. Thus you get a complete control over graphics objects. MATLAB's entire system of object-oriented graphics and its user controllability is referred to as 'Handle Graphics™'. Here we briefly discuss this system and its usage, but we urge the more interested reader to consult the Users Guide [1] for more details.

The key to understanding and using Handle Graphics system is to know how to get the handles of graphics objects and how to use handles to get and change properties of the objects. Since not all graphics objects are independent (for example, the appearance of a line depends on the current axes in use), and a certain property of one may affect the properties of the others, it is also important to know how the objects are related.

6.4.1 The object hierarchy

Graphics objects follow a hierarchy of parent-child relationship. The following tree diagram shows the hierarchy.

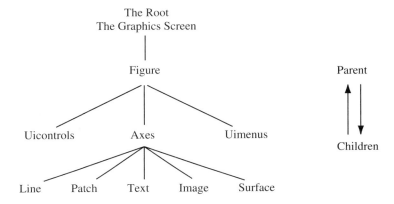

It is important to know this structure for two reasons:

- It shows you which objects will be affected if you change a default property value at a particular level, and
- It tells you at which level you can query for handles of which objects.

6.4.2 Object handles

Object handles are unique identifiers associated with each graphics object. These handles have a floating point representation. Handles are created at the time of creation of the object by graphics functions such as `plot(x,y)`, `contour(z)`, `line(z1,z2)`, and `text(xc,yc,'Look at this')`, etc.

Getting object handles

There are two ways of getting hold of handles:

1. By creating handles explicitly at the object-creation-level commands (that is, you can make a plot and get its handle at the same time):

   ```
   hl = plot(x,y,'r-')      returns the handle of the line to hl.
   hxl = xlabel('Angle')   returns the handle of the x-label to hxl.
   ```

2. By using explicit handle-returning functions:

   ```
   gcf                 gets the handle of the current figure.
   ```
   Example: `hfig = gcf;` returns the handle of the current figure in `hfig`.

gca   gets the handle of the current axes.

Example: haxes = gca; returns the handle of the current axes in haxes.

gco   gets the handle of the current object.

Handles of other objects, in turn, can be obtained with the get command. For example, hlines = get(gca,'children') returns the handles of all *children* of the current axes in a column vector hlines. The function get is used to get a property value of an object, specified by its handle, in the command form

get(*handle*,'*PropertyName*').

For an object with handle h, type get(h) to get a list of all property names and their current values.

Examples:

hl = plot(x,y)   plots a line and returns the handle hl of the plotted line.

get(hl)   lists all properties of the line and their values.

get(hl,'type')   shows the type of the object (e.g. line`

get(hl,'linestyle')   returns the current line-style of the line.

For more information on get, see the on-line help.

6.4.3 Object properties

Every graphics object on the screen has certain properties associated with it. For example, the properties of a line include type, parent, visible, color, linestyle, linewidth, xdata, ydata etc. Similarly, the properties of a text object, such as xlabel or title, include type, parent, visible, color, fontname, fontsize, fontweight, string, etc. Once the handle of an object is known, you can see the list of its properties and their current values with the command get(*handle*). For example, see Fig. 6.12 for the properties of a line and their current values.

There are some properties common to all graphics objects. These properties are: children, clipping, parent, type, userdata, and visible.

```
>> t=linspace(0,pi,50);
>> hL = line(t,sin(t));
>> get(hL);
 Color = [0 0 1]
 EraseMode = normal
 LineStyle = -
 LineWidth = [0.5]
 Marker = none
 MarkerSize = [6]
 MarkerEdgeColor = auto
 MarkerFaceColor = none
 XData = [ (1 by 50) double array]
 YData = [ (1 by 50) double array]
 ZData = []

 BeingDeleted = off
 ButtonDownFcn =
 Children = []
 Clipping = on
 CreateFcn =
 DeleteFcn =
 BusyAction = queue
 HandleVisibility = on
 HitTest = on
 Interruptible = on
 Parent = [3.00037]
 Selected = off
 SelectionHighlight = on
 Tag =
 Type = line
 UIContextmenu = []
 UserData = []
 Visible = on
```

Create a line with handle hL.

Query the line's properties and their current values with the `get` command.

Figure 6.12: Example of creating a line with an explicit handle and finding the properties of the line along with their current values.

Setting property values

You can see the list of properties and their values with the command `set(handle)`. Any property can be changed by the command

> set(*handle*, '*PropertyName*', *PropertyValue*')

where *PropertyValue* may be a character string or a number. If the *PropertyValue* is a string then it must be enclosed within single quotes.

Fig. 6.13 shows the properties and property-values of a line.

```
>> t=linspace(0,pi,50);              Create a line with handle hL.
>> x=t.*sin(t);
>> hL = line(t,x);                   Query the line's properties that
>> set(hL)                           can be set and the available
 Color                               options.
 EraseMode: [ {normal} | background | xor | none ]
 LineStyle: [ {-} | -- | : | -. | none ]
 LineWidth
 Marker: [ + | o | * | . | x | square | diamond ..
 MarkerSize
 MarkerEdgeColor: [ none | {auto} ] -or- a ColorSpec.
 MarkerFaceColor: [ {none} | auto ] -or- a ColorSpec.
 XData
 YData
 ZData

 ButtonDownFcn
 Children
 Clipping: [ {on} | off ]
 CreateFcn
 DeleteFcn
 BusyAction: [ {queue} | cancel ]
 .
 .
 .
```

Figure 6.13: Example of creating a line with an explicit handle and finding the properties of the line along with their possible values.

Now let us look at two examples:

Example-1: We create a line along with an explicit handle and then use the **set** command to change the line style, its thickness, and some of the data. See page 195.

Example-2: We write some text at a specified position (in the figure window), create its handle, and then use the **set** command to change the fontsize, font, and string of the text. See page 196.

Example Script	**Output**

Create a simple line plot and assign
its handle to hL.

```
t=linspace(0,pi,50);
x=t.*sin(t);
hL = line(t,x);
```

Change the line style to dashed.

```
set(hL,'linestyle','--')
```

Change the line thickness.

```
set(hL,'linewidth',3)
```

Change the values of some y-coordinates.

```
yvec=get(hL,'ydata');
yvec(25:35)=ones(size(yvec(25:35)));
set(hL,'ydata',yvec)
```

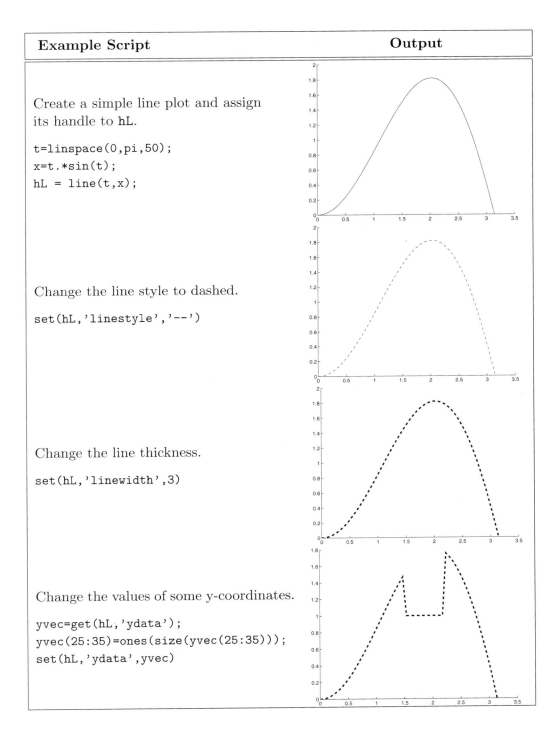

Example Script	Output

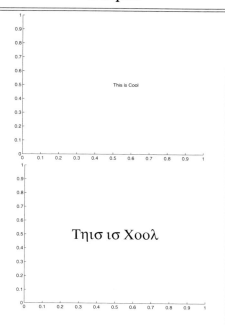

Write some text at the location (0.5,0.5)
and create a handle for it.

```
x=0.5; y=0.5;
hT = text(x,y,'This is Cool',...
    'erasemode','xor')
```

Make the text centered at (0.5,0.5) and
change the font and fontsize.

```
set(hT, 'horizontal','center','fontsize',36,...
    'fontname','symbol')
set(gca,'visible','off')
```

Now, create a template for presentation.

```
clf
line([0 0 1 1 0],[0 1 1 0 0]);
h1=text(.5,.7,'Coming Soon...',...
    'fontangle','italic',...
    'horizontal','center');
set(gca,'visible','off')
h2=text(.5,.5,'3D-Simulation',...
    'horizontal','center',...
    'fontsize',40,'fontname','times',...
    'fontweight','bold','erase','xor');
h3=text(.5,.4,'by','horizontal','center');
h4=text(.5,.3,'Stephen Spielberg',...
    'fontsize',16,'horizontal','center',...
    'erase','xor');
```

Next slide please...

```
set(h1,'string','')
set(h2,'string','The Model')
set(h3,'string','')
set(h4,'string','Assumptions & Idealizations')
```

6.4.4   Modifying an existing plot

Even if you create a plot without explicitly creating object handles, MATLAB creates handles for each object on the plot. If you want to modify any object, you have to first get its handle. Here is where you need to know the parent-child relationship among several graphics objects. The following example illustrates how to get the handle of different objects and use the handles to modify the plot.

We take Fig. 6.2 on page 164 and use the aforementioned Handle Graphics features to modify the figure. The following script file is used to change the plot in Fig. 6.2 to the one shown in Fig. 6.14. You may find the following script file confusing because it uses a vector of handles, **hline**, and accesses different elements of this vector, without much explanation. Hopefully, your confusion will be cleared after you read the next section *Understanding a vector of handles*. [Note: Before using the following commands, you must execute the commands shown in Fig. 6.2.]

```
h=gca;                           % get the handle of the current axes
set(h,'box','off');              % throw away the enclosing box frame
hline=get(h,'children');         % get the handles of children of axes
                                 %-- Note that hline is a vector of
                                 %-- handles because h has many  children
set(hline(7),'linewidth',4)      % change the line width of the 1st  line
set(hline(6),'visible','off')    % make the 'lin. approx' line  invisible
delete(hline(3))                 % delete the text 'linear  approximation'
hxl=get(h,'xlabel');             % get the handle of xlabel
set(hxl,'string','t (angle)')    % change the text of xlabel
set(hxl,'fontname','times')      % change the font of xlabel
set(hxl,'fontsize',20,'fontweight','bold')
                                 % change the font-size & font-weight
```

Understanding a vector of handles: In the above script file you may perhaps be confused about the use of the handle **hline**. The command, **hline = get(h,'children')**, above gets the handles of all the children of the current axes (specified by handle **h**) in a column vector **hline**. The vector **hline** has seven elements—three handles for the three lines and four handles for the four text objects (created by **text** and **gtext** commands). So, how do we know which handle is for which line or which text? The command **get(hline(i),'type')** lists the type of the object whose handle is **hline(i)**. The confusion is not clear yet. What if **hline(5)**, **hline(6)**, and **hline(7)** are all lines? How do we know which handle corresponds to which line? Once we know the type of the object, we can identify its handle, among several similar object handles, by querying a more distinctive property of the object, such as **linestyle** for lines and **string** for text objects. For example, consider the handle vector **hline** above. Then,

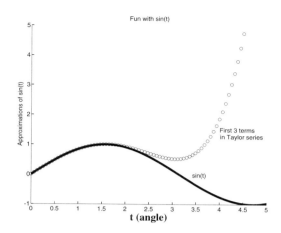

Figure 6.14: Example of manipulation of a figure with Handle Graphics. This figure is a result of executing the above script file after generating Fig. 5.1

get(hline(5),'marker')   returns 'o' for the linestyle,
get(hline(6),'linestyle')   returns '--' for the linestyle,
get(hline(1),'string')   returns 'in Taylor series' for the string,
get(hline(2),'string')   returns 'First 3 terms' for the string.

From this example, it should be clear that *the handles of children of the axes are listed in the stacking order of the objects*, i.e., the last object added goes on the top of the stack. Thus the elements of the handle vector correspond to the objects in the reverse order of their creation!

Deleting graphics objects

Any object in the graphics window can be deleted without disturbing the other objects with the command

delete(ObjHandle)

where *ObjHandle* is the handle of the object. We have used this command in the script file which produced Fig. 6.14 to delete the text 'linear approximation' from the figure. We could have used delete(hline(6)) to delete the corresponding line rather than making it invisible.

Modifying plots with PropEdit

Now, that you have some understanding of Handle Graphics, object handles, and object properties, you may like to use the point-and-click graphics editor, *PropEdit*. Simply type propedit to activate the editor. All graphics objects from the active figure window are shown along with their properties in the

PropEdit window. You can select a property from the list by clicking on it and then change it in the narrow rectangle in the middle. A graphics object with a plus (+) on its left indicates that you can double click on it to see its children.

6.4.5 Complete control over the graphics layout

We close this section with an example of arbitrary placement of axes and figures in the graphics window. With Handle Graphics tools like these, you have almost complete control of the graphics layout. Here are two examples.

Example-1: Placing insets

The following script file shows how to create multiple axes, size them and place them so that they look like insets. The output appears in Fig. 6.15

```
%-------------------------------------------------------------------
%        Example of graphics placement with Handle Graphics
%-------------------------------------------------------------------
clf                               % clear all previous figures
t=linspace(0,2*pi);  t(1) = eps;  % t(1) is set to a small number
y=sin(t);
%--------------------
h1=axes('position',[0.1 0.1 .8 .8]);% place axes with width .8 and
                                  %- height .8 at coordinates (.1,.1)
plot(t,y),xlabel('t'),ylabel('sin t')
set(h1,'Box','Off');              % Turn the enclosing box off
xhl=get(gca,'xlabel');            % get the handle of 'xlabel' of the
                                  %- current axes and assign to  xhl
set(xhl,'fontsize',16,'fontweight','bold')
                                  % change attributes of 'xlabel'
yhl=get(gca,'ylabel');            % do the same with 'ylabel'
set(yhl,'fontsize',16,'fontweight','bold')
%--------------------
h2=axes('position',[0.6 0.6 .2 .2]);% place another axes on the same plot
fill(t,y.^2,'r')                  % draw a filled polygon with red fill
set(h2,'Box','Off');
xlabel('t'),ylabel('(sin t)^2')
set(get(h2,'xlabel'),'FontName','Times')
set(get(h2,'ylabel'),'FontName','Times')
%--------------------
h3=axes('position',[0.15 0.2 .3 .3]); % place yet another axes
polar(t,y./t);                    % make a polar plot
polarch=get(gca,'children');      % get the handle of all the  children
                                  %- of the current axes
set(polarch(1),'linewidth',3)     % set the line width of the  first child
                                  %- which is actually the line  we plotted
```

```
for i=1:length(polarch)              % clear the clutter due to text
   if strcmp(get(polarch(i),'type'),'text')  % look for 'text' objects
      delete(polarch(i))             % delete all text objects
   end
end
%----------------------
```

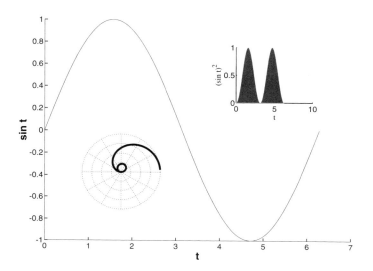

Figure 6.15: Example of manipulation of the Figure Window with Handle Graphics. Virtually anything in the figure window, including the placement of axes, can be manipulated with Handle Graphics.

Example-2: Fun with spirals

So, now that you know how to create an **axes**, position it, and make plots in it, let us have some fun. How about doing some artwork with MATLAB? Place and size the axes, take the outer box off at will, change the color of the axes, put the x-axis on top, y-axis on the right and so on. Go ahead, try the following script file. It produces the spirals that appear in Fig. 6.16 and, in color, on the cover page of this book.

```
% FunWithSpirals: Script to plot 4 filled spirals in different axes
% Written by Rudra Pratap on July 7, 1997,
%             last modified June 25, 1998.
% ------------------------
t = linspace(0,8*pi,200);   % create basic data for a spiral
r=exp(-.1*t);
x=r.*cos(t);
```

```
y=r.*sin(t);

clf                          % clear previous figure settings
h1 = axes('position',[.1,.1,.5,.5]);      % first axes
fill(x,y,'g')                % first (big) spiral in green
h2 = axes('position',[.45,.45,.3,.3]);    % second axes
fill(x,y,'b')                % second spiral in blue
set(h2,'xcolor','b');        % change x-axis color to blue
set(h2,'ycolor','b');        % change y-axis color to blue
set(h2,'xticklabel',' ');    % remove axis tick marks
set(h2,'yticklabel',' ');
h3 = axes('position',[.67,.67,.2,.2]);    % third axes
fill(x,y,'m'), box('off')    % third spiral, no outer box
set(h3,'xcolor','m');        % change axis color to magenta
set(h3,'ycolor','m');
set(h3,'xticklabel',' ');    % remove axis tick marks
set(h3,'yticklabel',' ');
h4 = axes('position',[.84,.84,.1,.1]);    % fourth axes
fill(x,y,'r')                % fourth spiral in red
set(h4,'color','y');         % change background color
set(h4,'xaxisloc','top');    % locate x-axis on top
set(h4,'yaxisloc','right');  % locate y-axis on right
```

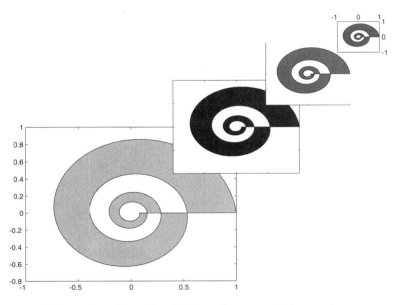

Figure 6.16: Example of manipulation of axes and its various properties.

For on-line help
type:
`help graphics`

6.5   Saving and Printing Graphs

The simplest way to get a hardcopy of a graph is to type `print` in the command window after the graph appears in the figure window. The `print` command sends the graph in the current figure window to the default printer in an appropriate form. On PCs (running Windows) and Macs you could, alternatively, activate the figure window (bring to the front by clicking on it) and then select `print` from the file menu.

The figure can also be saved into a specified file in the PostScript or Encapsulated PostScript (EPS) format. These formats are available for black and white as well as color printers. The PostScript supported includes both Level 1 and Level 2 PostScript. The command to save graphics to a file has the form

> `print -ddevicetype -options filename`

where *devicetype* for PostScript printers can be one of the following:

devicetype	Description	*devicetype*	Description
ps	black and white PostScript	eps	black and white EPSF
psc	color PostScript	epsc	color EPSF
ps2	Level2 BW PostScript	eps2	Level 2 black and white EPSF
psc2	Level 2 color PostScript	epsc	Level 2 color EPSF.

For example, the command

> `print -deps sineplot`

saves the current figure in the Encapsulated PostScript file `sineplot.eps`. The '.eps' extension is automatically generated by MATLAB.

The standard optional argument *-options* supported are `append`, `epsi`, `Pprinter`, and `fhandle`. There are several other platform dependent options. See the on-line help on `print` for more information.

In addition to the PostScript devices, MATLAB supports a number of other printer devices on UNIX and PC systems. There are device options available for HP Laser Jet, Desk Jet, and Paint Jet printers, DEC LN03 printer, Epson printers and other types of printers. See the on-line help on `print` to check the available devices and options.

Other than printer devices, MATLAB can also generate a graphics file in the following popular formats among others.

`-dill`	saves file in Adobe Illustrator format
`-djpeg`	saves file as a JPEG image
`-dtiff`	saves file as a compressed TIFF image
`-dmfile`	saves file as an M-file with graphics handles

It is also possible to save a graphics file as a list of commands (an M-file), regenerate the graphics later, and modify it. To save the graphics in the currently active window, type:

`print `*`filename`*` -dmfile`

Later, to get the plot again, simply execute the file you just created.

The Adobe Illustrator format is quite useful if you want to dress-up or modify the figure in a way that is very difficult to do in MATLAB. Of course, you must have access to Adobe Illustrator to be able to open and edit the saved graphs. Figure 6.17 shows an example of a graph generated in MATLAB and then modified in Adobe Illustrator. The most annoying aspect of MATLAB 4.x's graphics was the lack of simple facilities to write subscripts and superscripts and mix fonts in the labels. MATLAB 5 has solved this problem by incorporating a subset of LaTeXcommands which can be used in labels and texts. If you do not know LaTeX, it may be worth browsing the MATLAB `helpdesk` to see the list of commands incorporated and their output.

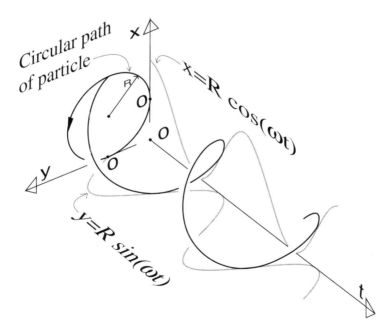

Figure 6.17: Example of a figure generated in MATLAB, saved in the Illustrator format and then modified in Adobe Illustrator. The rotation and shearing of texts was done in Illustrator (courtesy A. Ruina).

6.6   Animation

We all know the visual impact of animation. If you have a lot of data representing a function or a system at several time sequences, you may wish to take advantage of MATLAB's capability to animate your data.

There are three types of facilities for animation in MATLAB.

1. **Comet plot:** This is the simplest and the most restricted facility to display a 2-D or 3-D line graph as an animated plot. The command `comet(x,y)` plots the data in vectors x and y with a comet moving through the data points. The trail of the comet traces a line connecting the data points. So, rather than having the entire plot appear on the screen at once, you can see the graph 'being plotted'. This facility may be useful in visualizing trajectories in a phase plane. For an example, see the built-in demo on the Lorenz attractor.

2. **Movies:** If you have a sequence of plots that you would like to animate, use the built-in `movie` facility. The basic idea is to store each figure as a frame of the movie, with each frame stored as a column vector of a big matrix, say M, and then to play the frames on the screen with the command `movie(M)`. A frame is stored in a column vector using the command `getframe`. For efficient storage you should first initialize the matrix M. The built-in command `moviein` is provided precisely for this initialization, although you can do it yourself too. An example script file to make a movie might look like this:

```
%--- skeleton of a script file to generate and play a movie  ---
%
nframes = 36;              % number of frames in the the movie
Frames = moviein(nframes); % initialize the matrix 'Frames'
for i = 1:nframes
   :                       % you may have calculations here to
   :                       %- generate data
   :
   x = ....;
   y = ....;
   plot(x,y)              % you may use any plotting function
   Frames(:,i) = getframe; % store the current figure as a frame
end
movie(Frames,5)           % play the movie Frames 5 times
```

You can also specify the speed (frames/second) at which you want to play the movie (the actual speed will eventually depend on your CPU) by typing `movie(Frames,`$m,$ *fps*`)`, which plays the movie, stored in **Frames,** m times at the rate *fps* frames per second.

3. **Handle Graphics:** Another way, and perhaps the most versatile way, of creating animation is to use the Handle Graphics facilities. The basic idea here is to plot an object on the screen, get its handle, use the handle to change the desired properties of the object (most likely its 'xdata' and 'ydata' values), and replot the object over a selected sequence of times. There are two important things to know to be able to create animation using Handle Graphics:

- The command **drawnow**, which flushes the graphics output to the screen without waiting for the control to return to MATLAB. The on-line help on **drawnow** explains how it works.
- The object property 'erasemode' which can be set to 'normal', 'background', 'none', or 'xor' to control the appearance of the object when the graphics screen is redrawn. For example, if a script file containing the following lines is executed

```
h1 = plot(x1,y1,'erasemode','none');
h2 = plot(x2,y2,'erasemode','xor');
   :
newx1=...; newy1=...; newx2=...; newy2=...;
   :
set(h1,'xdata',newx1,'ydata',newy1);
set(h2,'xdata',newx2,'ydata',newy2);
   :
```

then the first **set** command draws the first object with the new x-data and y-data, but the same object drawn before remains on the screen, while the second **set** command redraws the second object with new data and also erases the object drawn with the previous data x2 and y2. Thus it is possible to keep some objects fixed on the screen while some other objects change with each pass of a control flow.

Now, let us look at some examples that illustrate the use of Handle Graphics in animation.

Example-1: A bead goes around a circular path: The basic idea is to first calculate various positions of the bead along the circular path, draw the bead as a point at the initial position and create its handle, and then use the handle to set the x- and y-coordinates of the bead to new values inside a loop that cycles through all positions. The **erasemode** property of the bead is set to **xor** (Exclusive Or) so that the old bead is erased from the screen when the new bead is drawn. Try the following script file.

```
% Script file for animating the circular motion of a bead
% ---------------------------------------------
clf                                     % clear any previous figure
theta=linspace(0,2*pi,100);             % create a vector theta
x=cos(theta);                           % generate x and y-coordinates
y=sin(theta);                           %- of the bead along the path
hbead=line(x(1),y(1),'marker','o',...
    'markersize',8,'erase','xor');      % draw the bead at the initial
                                        %- position and assign a handle
axis([-1 1 -1 1]); axis('square');
for k=2:length(theta)                   % cycle through all positions
    set(hbead,'xdata',x(k),'ydata',y(k));
                                        % draw the bead at the new position
    drawnow
end
```

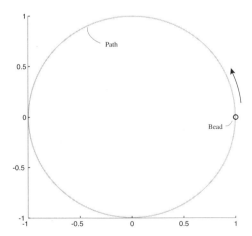

Figure 6.18: A bead goes on a circular path.

Example-2: The bead going around a circular path leaves its trail:
In Example-1 above, the bead goes on the circular path, but it does not
clearly seem that it traverses a circle. To make it clear, we can make the
bead leave a trail as it moves. For this purpose, we basically draw the bead
twice, once as a bead (with bigger marker size) and once as a point at each
location. But we set the **erasemode** property of the point to **none** so that
the point (the previous position of the bead) remains on the screen as the
bead moves and thus creates a trail of the bead.

```
% Script file for animating the circular motion of a bead. As the
% bead moves, it leaves a trail behind it.
% ------------------------------------------
clf
theta=linspace(0,2*pi,100);
x=cos(theta); y=sin(theta);
hbead=line(x(1),y(1),'marker','o','markersize',8,'erase','xor');
htrail=line(x(1),y(1),'marker','.','color','r','erase','none');
axis([-1 1 -1 1]);
axis('square');
for k=2:length(theta)
    set(hbead,'xdata',x(k),'ydata',y(k));
    set(htrail,'xdata',x(k),'ydata',y(k));
    drawnow
end
```

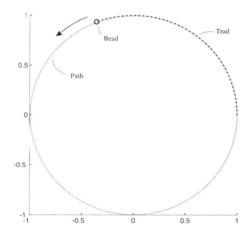

Figure 6.19: A bead goes on a circular path and leaves a trail behind it.

Example-3: A bar pendulum swings in 2D: Here is a slightly more complicated example. It involves animation of the motion of a bar pendulum governed by the ODE $\ddot{\theta} + \sin\theta = 0$. Now that you are comfortable with defining graphics objects and using their handles to change their position etc., the added complication of solving a differential equation should not be too hard.

```
%----- script file to animate a bar pendulum --------------

    clf                          % clear figure and stuff
    data=[0 0; -1.5 0];          % coordinates of endpoints of the bar
    phi=0;                       % initial orientation
    R=[cos(phi) -sin(phi);  sin(phi) cos(phi)];
                                 % rotation matrix
    data=R*data;
    axis([-2 2 -2 2])            % set axis limits
    axis('equal')

    %-----define the objects called bar, hinge, and path.
    bar=line('xdata',data(1,:),'ydata',data(2,:),...
             'linewidth',3,'erase','xor') ;
    hinge=line('xdata',0,'ydata',0,'marker','o','markersize',[10]);
    path=line('xdata',[],'ydata', [],'marker','.','erasemode','none');

    theta=pi-pi/1000;            % initial angle
    thetadot=0;                  % initial angular speed
    dt=.2; tfinal=50; t=0;       % time step, initial and final time

    %------Euler's method for numerical integration
    while(t<tfinal);
        t=t+dt;
        theta=theta + thetadot*dt;
        thetadot=thetadot -sin(theta)*dt;
        R=[cos(theta)  -sin(theta); sin(theta) cos(theta)];
        datanew= R*data;

        %---- change the property values of the objects: path and bar.
        set(path,'xdata', datanew(1,1), 'ydata', datanew(2,1) );
        set(bar,'xdata',datanew(1,:),'ydata',datanew(2,:) );
        drawnow;
    end
```

Example-4: The bar pendulum swings, and other data are displayed: Now here is the challenge. If you can understand the following script file, you are in good shape! You are ready to do almost any animation. The following example divides the graphics screen in four parts, shows the motion of the pendulum in one part, shows the position of the tip in the second part, plots the angular displacement θ in the third part and the angular speed $\dot{\theta}$ in the fourth part (see Fig. 6.20). There are four animations occurring simultaneously. Try it! There is an intentional bug in one of the four animations. If you start the pendulum from the vertical upright position with an initial angular speed (you will need to change `thetadot = 0` statement inside the program for this) then you should see the bug. Go ahead, find the bug and fix it.

```
%----- script file to animate a bar pendulum and the data --------

% get basic data for animation
% ask the user for initial position
disp('Please specify the initial angle from the')
disp('vertical upright position.')
disp(' ')
offset = input('Enter the initial angle now: ');
                        % ask the user for time of simulation
tfinal = input('Please enter the duration of simulation: ');
disp('I am working....')

theta = pi-offset;       % initial angle
thetadot = 0;            % initial angular speed
dt = .2; t=0;  tf=tfinal; % time step, initial and final time

disp('Watch the graphics screen')
clf                      % clear figure and stuff
h1 = axes('position',[0.6 .1 .4 .3]);
axis([0 tf -4 4]);       % set axis limits
xlabel('time'),ylabel('displacement')
Displ = line('xdata',[],'ydata', [],'marker','.','erasemode','none');

h2 = axes('position',[0.6 .55 .4 .3]);
axis([0 tf -4 4]);       % set axis limits
xlabel('time'),ylabel('velocity')
Vel = line('xdata',[],'ydata', [],'marker','.','erasemode','none');

h3 = axes('position',[.1 .1 .4 .4]);
axis([-pi pi -4 4])      % set axis limits
axis('square')
Phase = line('xdata',[],'ydata', [],'marker','o',...
          'markersize',5,'erasemode','none');
h4 = axes('position',[.1 .55 .4 .4]);
```

```
axis([-2 2 -2 2])          % set axis limits
axis('square')

data = [0 0; -1.8 0];      % coordinates of endpoints of the bar
phi = 0;                   % initial orientation
R = [cos(phi) -sin(phi); +sin(phi) cos(phi)];   % rotation matrix
data=R*data;
%-----define the objects called bar, hinge, and path.
bar = line('xdata',data(1,:),'ydata',data(2,:),...
          'linewidth',3,'erase','xor');
hinge = line('xdata',0,'ydata',0,'marker','o','markersize',[10]);
path = line('xdata',[],'ydata', [],'marker','.','erasemode','none');
%------Euler's method for numerical integration
while(t<tfinal);
    t = t+dt;
    theta = theta + thetadot*dt;
    thetadot = thetadot -sin(theta)*dt;
    R = [cos(theta) (-sin(theta)); sin(theta) cos(theta)];
    datanew = R*data;
    %---- change the property values of the objects: path and bar.
    set(path,'xdata', datanew(1,1), 'ydata', datanew(2,1) );
    set(bar,'xdata',datanew(1,:),'ydata',datanew(2,:) );
    set(Phase,'xdata', theta, 'ydata', thetadot);
    set(Displ,'xdata', t, 'ydata', theta );
    set(Vel,'xdata', t, 'ydata', thetadot );
    drawnow;
end
```

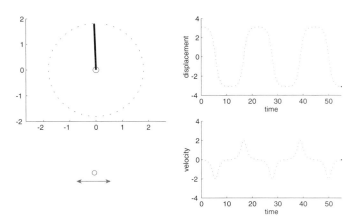

Figure 6.20: Animation of motion of a bar pendulum along with animation of position and velocity data.

7. *Errors*

Errors are an integral part of life whether you interact with computers or not. The only difference is, when you interact with computers your errors are pointed out immediately—often bluntly and without much advice. Interaction with MATLAB is no exception. Yes, to err is human, but to forgive is definitely not MATLABine. So, the earlier you get used to the blunt manners of your friend and his terse comments, the better for you. Since this friend does not offer much advice most of the time, we give you some hints here based on our own experience of dealing with your friend. Before we begin, we warn you that this friend has a tendency of becoming very irritating if you work under too much time pressure or don't have enough sleep. In particular, if you are not relaxed enough to distinguish between '(' and '[', ';' and ':', or 'a' and 'A', you and your friend are going to have long sessions staring at each other.

Here are the most common error messages, in alphabetical order (according to the first word that is not an article or a preposition), that you are likely to get while working in MATLAB. All messages below are shown following a typical command. Following the actual message are explanations and tips.

- ```
  >> D(2:3,:)=sin(d)
  ???  In an assignment A(matrix,:) = B, the number
  of columns in A and B must be the same.
  ```
  This is a typical problem in matrix assignments where the dimensions of the matrices on the two sides of the equal sign do not match. Use the `size` and `length` commands to check the dimensions on both sides and make sure they agree. For example, for the above command to execute properly, `size(D(2:3,:)` and `size(sin(d))` or `size(d)` must give the same dimensions.
  ```

A similar error occurs when trying to assign a matrix to a vector:

```
>> D(:,2)=d1
??? In an assignment  A(matrix) = B, a vector A
can't be resized to a matrix.
```

In this example, D and d1 are matrices, but D(:,2) is a vector (the second column of D), so d1 cannot fit into D(:,2).

- ```
  >> (x,y)=circlefn(5);
  ??? (x,
         |
  Error: ")" expected, "," found.
  ```

  Not really! But MATLAB is confused with the list of variables because a list within *parenthesis* represents matrix indices. When the variables represent output of a function or a list of vectors, they must be enclosed within *square brackets*. The correct command here is [x,y]=circlefn(5) (circlefn is a user-written function). When parentheses and brackets are mixed up, the same error message is displayed:

  ```
  (x,y]=circlefn(5);
  ??? (x,
          |
  Error: ")" expected, "," found.
  ```

- ```
  x=1:10;
  v=[0 3 6];
  x(v)
  ??? Index into matrix is negative or zero.
  ```

  The first element of the index vector v is zero. Thus we are trying to get the zeroth element of x. But zero is not a valid index for any matrix or vector in MATLAB. The same problem arises when a negative number appears as an index. Also, of course, an error occurs when the specified index exceeds the corresponding dimension of the variable:

  ```
  x(12)
  ??? Index exceeds matrix dimensions.
  ```

  The examples given here for index-dimension mismatch are almost trivial. Most of the times these problems arise when matrix indices are created, incremented, and manipulated inside loops.

- ```
  >> x=1:10; y=10:-2:-8;
  >> x*y
  ??? Error using ==> *
  Inner matrix dimensions must agree.
  ```
  ```

In matrix multiplication x*y the number of rows of x must equal the number of columns of y. Here x and y are both row vectors of size 1×10 and therefore, cannot multiply. However, x*y' and x'*y will both execute without error, producing inner and outer products respectively. Several other operations involving matrices of improper dimensions produce similar errors. A general rule is to write the expression on paper and think if the operation makes mathematical sense. If not, then MATLAB is likely to give you error. For example, A^2 makes sense for a matrix A only if the matrix is square, and A^x for a vector x and matrix A does not make any sense. The exceptions to this rule are the two division operators—'/' and '\'. While y/x, for the two vectors defined above, may not make any sense mathematically, MATLAB gives an answer:

```
>> y/x
ans =
   -0.2857
```

This is because this division is not just division in MATLAB, but it also gives solutions to matrix equations. For rectangular matrices it gives solutions in the least squares sense. See on-line help on **slash** for more details.

A common source of error is to use the matrix operators where you want array operators. For example, for the vectors x and y above, y.^x gives element-by-element exponentiation, but y^x produces an error:

```
>> y^x
??? Error using ==> ^
Matrix dimensions must agree.
```

- ```
  [x,y]=circlefn;
  ??? Input argument r is undefined.

  Error in ==> D:\MATLAB_files\circlefn.m
  On line 4   ==> x = r*cos(theta); y = ....
  ```
  A function file has been executed without giving proper input. This is one of the very few error messages that provides enough information (the function name, the name of the directory where the function file is located, and the line number where the error occurred).

- ```
  >> [t,x]=Circle(5);
  ??? Attempt to execute SCRIPT Circle as a function.
  ```

  Here **Circle** is a script file. Input-output lists cannot be specified with script files. But here is a slightly more interesting case that produces the same error. The error occurs in trying to execute the following function file:

```
Function [x,y] = circlefn(r);
% CIRCLEFN - Function to draw a circle of radius r.
theta = linspace(0,2*pi,100);        % create vector theta
x = r*cos(theta); y = r*cos(theta); % generate coordinates
plot(x,y);                            % plot the circle
```

Here is the error:

```
>> [x,y]=circlefn(5)
??? Attempt to execute SCRIPT circlefn as a function.
```

You scream, "Hey, it's not a script!" True, but it is not a function either. For it to qualify as a function, the f in function, in the definition line, must also be in lower case which is not the case here. So MATLAB gets confused. Yes, the error message could have been better, but you are probably going to say this in quite a few cases.

- ```
  [x,y]=Circlefn();
  ??? [x,y]=Circlefn()
                     |
  ```

  ```
  Error: Expected a variable, function, or constant, found "(
  ```
  This is an obvious one—the input variable is missing. MATLAB expects to see a variable or a function name within the parentheses.

- ```
  CIRCLEFN[5];
  ??? CIRCLEFN[
               |
  ```

  ```
  Error: Missing operator, comma, or semi-colon.
  ```
  Here parentheses are required in the place of the square brackets. The error locator bar is at the right place and should help. But what is missing is neither an *operator* nor a *comma* nor a *semi-colon*. This error message is perhaps the most frequent one and also the most useless one in terms of helping you out.

```
circle(5);
??? Attempt to execute SCRIPT Circle as a function.
```

circle is a script file and cannot take an input as an argument. Again, the error message does not help much.

```
EIG(D)
??? Capitalized internal function EIG; caps lock may be on.
```

Here EIG is a built-in function and its name must be typed in lower case: eig. MATLAB does not recognize EIG, and hence the error. But, this time, the error message is quite helpful.

- >> x = b+2.33
  ??? Undefined function or variable b.

  The variable b has not been defined. This message is right on target. But when the same message comes for a function or script which you have written, you may scratch your head. In such cases, the function or the script that you are trying to execute is most probably in a different directory than the current directory. Use what, dir, or ls to show the list of files in the current directory. If the file is not listed there then MATLAB cannot access it. You may have to locate the directory of the file with the command which *filename* and then change the working directory with the cd command to the desired directory.

- >> global a, b
  ??? Undefined function or variable b.

  You think that you are merely declaring a and b to be global variables. But, what is that comma doing there (after a)? MATLAB treats the comma to be a command or statement separator. Thus it did accept the declaration global a as a valid command; now it is looking for a value of b (since typing a variable by itself directs MATLAB to return its value). So, do not use commas to separate variables in global, save, or load commands. A really tricky situation arises when you make a similar mistake in a *for*-loop statement: for i=1, n. This statement will execute without any error message, but the loop will be executed for only $i = 1$ and MATLAB will happily display the value of n on the screen.

- plot(d,d1)
  ??? Error using ==> plot
  Vectors must be the same lengths.

  The input vectors in plot command must be pair-wise compatible. For a detailed discussion of this, see the description of the plot command in Chapter 6.

8. *What Else is There?*

8.1 The Symbolic Math Toolbox

The Symbolic Toolbox allows MATLAB to respond to Maple commands. MATLAB is a program that does mostly arithmetic and Maple is another program that does mostly algebra and calculus. MATLAB mostly gives output that is a number or array of numbers. Maple is designed to give output in symbolic form. Matlab can tell you that $\sqrt{2}$ is about 1.41415 and Maple can tell you that the solutions to $x^2 = c$ are $x = \pm\sqrt{c}$ and leave it at that with the letter "c". MATLAB comes from the numeric calculation tradition of Fortran, Basic, and C. Maple comes from the artificial intelligence tradition of (first and some say best) Macsyma, later imitated by Mathematica.

Recently Macsyma, Maple, and Mathematica have added the ability to give numeric solutions and make nice plots. Even though these programs run slower, are harder to learn, and harder to use than MATLAB for these purposes Mathworks has felt the competition. So they responded. Mathworks found a way for MATLAB users to do symbolic math. How? Mathworks got a license to sell Maple as part of MATLAB. Much of Maple is in the "Symbolic Math Toolbox". For the rest of Maple (for really hard math) you need the Extended Symbolic Math Toolbox (more \$).

To master the Symbolic Math Toolbox you have to master Maple which is at least as much work as to learn MATLAB. You also have to learn how to run Maple from inside MATLAB and keep your head straight at the same time. This was difficult in MATLAB 4, but with MATLAB 5 onwards the situation has improved. Basically you type a command that makes sense to either plain MATLAB or Maple, and if plain MATLAB can't make sense of what you are requesting it checks to see if you have typed a legitimate Maple command (MATLAB's version of the Maple command, that is).

8.1.1 Should you buy it?

Although some people swear by them, symbolic calculations seem to be less useful than numeric computation to many people for two reasons. First, many problems are too hard to do symbolically. An infinitely fast computer doing symbolic computation will never tell you a formula for the integral of $\sin(\log(x)^2)$, but MATLAB will tell you the area under the curve as accurately as you would like (see Section 5.4, page 135). Second, the symbolic formulas one obtains are often unwieldy. How does, say, a 50 term symbolic formula give one more insight than a numeric calculation?

With the Symbolic Toolbox, MATLAB can do essentially all that Mathematica and Macsyma can do. Since the Symbolic Toolbox *comes* with the student edition of MATLAB, you might as well use Maple there for free. If you don't have the Symbolic Toolbox and you don't know of symbolic calculations that you need to do, wait. Most engineers can do fine without it. If you are one of the few that do *a lot* of involved symbolic calculation you might find Macsyma, Mathematica, or Maple more convenient as stand-alone products rather than trying to untangle Maple through the Symbolic Tool Box.

If you want to do a lot of back and forth between symbolic and numeric computation, then MATLAB's Symbolic Toolbox is just the thing for you.

8.1.2 Two useful tools in the Symbolic Math Toolbox

Before getting more involved, you should note two cute and useful tools that come with the symbolic math toolbox.

1. **A quick way to make plots: ezplot.** Here is the simplest way to make simple plots.

   - The command `ezplot sin(t)` makes a plot of the sin function on the default interval $(-2\pi, 2\pi)$. More elaborately you can also type `ezplot('sin(t)')`.
   - The command `ezplot t^2/exp(t)  [-1 5]` plots the function t^2/e^t over the interval $(-1, 5)$. More elaborately you can type `ezplot( 't^2/exp(t)', [-1 5])`.

   If you find you want more control of your plots than `ezplot` gives you, learn more of MATLAB's plotting features on page 29 and in Chapter 6.

2. **A fun tool: funtool.** Type `funtool` and you will be operating a two-screen plotting calculator that does symbolic calculations. Use this when you need to do some quick checking about a function, its derivative, integral, inverse, etc. The 'help' key explains what the other keys do. Some correspond to basic Maple commands that you can use at the MATLAB command line also.

Numeric vs symbolic computation

What is the difference between numeric (plain MATLAB) and symbolic (the Symbolic Math Toolbox) computation? Let us say, you wanted to know the derivative of a function, $t^2 \sin(3t^{1/4})$.

The definition of derivative gives a crude way to *numerically* calculate the derivative of a function as $f' \approx (f(x + h) - f(x))/h)$ if h is small. The MATLAB commands below calculate this approximation of the derivative at a set of points spaced $\Delta t = h$ apart.

```
h =.1;                      % delta t for the  difference
t = -pi : h : pi;           % the region of interest for t
f = t.^2.*sin(3*t.^(1/4));  % the function at all the t values
fprime = diff(f)/h;         % numerical approx of derivative
              % Note: diff(f) has 1 less element than f
plot( t(1:end-1),  fprime)  % plot of the derivative
```

The derivative of the function is represented by the two lists of numbers t and fprime. The derivative at t(7) is fprime(7).

Compare this with the Symbolic Toolbox calculation of the derivative of the same function. Here is the command and response.

```
>> symb_deriv=diff( 't^2 * sin(3*t^(1/4))' )

symb_deriv =

2*t*sin(3*t^(1/4))+3/4*t^(5/4)*cos(3*t^(1/4))
```

The Symbolic Toolbox gives you a formula for the derivative. If you are patient you can verify by hand that indeed

$$\frac{d}{dt}\left(t^2 \sin(3\,t^{1/4})\right) = 2\,t\sin(3\,\sqrt[4]{t}) + (3/4)\,t^{5/4}\cos(3\,\sqrt[4]{t}).$$

A plot of the function above gives about the same curve as the previous numeric calculation. If you type int(symb_deriv) you will get back, surprise, t^2*sin(3*t^(1/4)) as expected by the fundamental theorem of calculus. Maple is pretty good at basic calculus.

Notice that diff is two different commands. One in MATLAB (dealing with differences between consecutive items in a list) and one in Maple (symbolically calculating the derivative). Whether MATLAB or Maple responds when you use diff depends on whether you have typed something in the correct syntax for one or the other program.

8.1.3   Getting help with the Symbolic Toolbox

There are many books written about Maple. These few pages are the glimmer of the tip of an iceberg. To go further without going to the library or bookstore, you can get help from MATLAB's usual labyrinth of help options.

- If you know the name of the command you want help with, say `solve` you can see helpful explanations any of three ways:

  1. type `help solve` at the command line. Luckily `solve` is a Maple command and is not also a plain MATLAB command. Beware, some commands have meaning to both plain MATLAB and the Symbolic Toolbox like the command `diff` used above. If you type `help diff` you will see a description of the plain old MATLAB command `diff` with this helpful clue at the end:

     ```
     Overloaded methods
         help sym/diff.m
         help char/diff.m
     ```

     "Overloaded" means the command "`diff`" has meaning outside of plain MATLAB. "`help sym/diff.m`" means that if you type `help sym/diff` you will get help with the Symbolic Toolbox (i.e., Maple) command also called `diff`.

  2. type `help sym/solve` at the command line to get the same help file as `help solve`. Type `help sym/diff` to get help on the *symbolic* command `diff`.

  3. Type `helpdesk` at the MATLAB prompt. Then click on <u>Symbolic Math Toolbox Ref</u> and you will get an alphabetical list of underlined commands. Click on any one for more information, a little different from the help file but the same as in the second 100 pages or so of the MATLAB Symbolic Math Toolbox manual.

- To see an organized list of Maple commands with a very short description you can do either of two things:

  1. type `help symbolic` on the command line. One line in this list is, for example,
     "`solve - Symbolic solution of algebraic equations.`".

  2. Click on the ? button at the top of the command window. Then click on the line "`Toolbox:symbolic - Symbolic Math Toolbox`" and you will be presented with the same list but in a separate help window.

- "Live" demonstrations of a few groups of commands are available. These demos take about a minute if you gloss your eyes and repeatedly hit the space bar. They take 10-30 minutes if you try to follow them carefully. The demos are: `symintro` (introduction), `symcalcdemo` (calculus), `symlindemo` (linear algebra), `symvpademo` (variable precision arithmetic), `symrotdemo` (rotations in the plane), `symeqndemo` (solution of equations). There are two ways to get to these demos:

1. Type, say, `symcalcdemo` on the command line and then do as told.

2. Type `demos` at the command line. Then click on **Symbolic Math**, then click on, say, **Introduction**, and then click on the topic of your choice.

- The 250 page *Symbolic Math Toolbox* manual is on your computer if someone installed it. At the command line type `helpdesk`. In the WWW browser that pops up click on <u>Full Documentation Set</u>. Then find and click on Symbolic <u>Math Toolbox User's Guide</u>. You are then brought to the program Adobe Acrobat which lets you find your way through the manual. The manual is not as good as many introductory books on Maple.

8.1.4 Using the Symbolic Math Toolbox

Lets see some other things the Symbolic Math Toolbox can do, besides `diff`, `int`, `ezplot`, and `funtool`.

1. **Basic manipulations:** Expand a polynomial, make it look nice, solve it, and check the solution. Try typing the lines below (one at a time or in an m-file) and keep track of MATLAB's response.

```
syms x   a                    % tell matlab that x and a are symbols
f       = (x-1) * (x-a) * (x + pi) * (x+2) *(x+3)    % define f
g       = expand(f)           % rewrite f, multiplying everything out
h       = collect(g)          % rewrite again by collecting terms
soln   = solve(h,x)           % find all the solutions,
check = subs(f,x,soln(5))  % check, say, the fifth solution
```

Comments:

- The `syms` declaration can sometimes be skipped. MATLAB can sometimes figure out by context that you want a letter to be a symbol. Two ways to do this are with the `sym` command and with single quotes ('). But it is safest to be explicit and use `syms`

- Since `x` is a symbol, `f` is automatically treated as a symbolic expression.

- `solve` is a powerful command that tries all kinds of things to find a solution. Here `solve` manages to find all five roots of a fifth order polynomial (something that cannot always be done, by the way).

- subs is an often used command if you do math online. Here every occurrence of x in the expression f is replaced with the fifth supposed solution. The result of this line of calculation is, predictably, 0.

2. **Talking to itself:** Get plain MATLAB to understand the output of the Symbolic Toolbox. One confusion is that the output of symbolic commands are symbolic expressions. These often *look exactly* like regular MATLAB expressions but MATLAB doesn't see them that way. There are a few tricks to getting symbolic expressions into a form that you can easily use with plain MATLAB. The key commands are double (turns a symbolic array of numbers into plain old numbers), eval (takes text that looks like a MATLAB command and executes the command), and vectorize (re-writes a formula so that it can be applied to a whole array of numbers). Try the commands below.

```
syms x t y a              % Set x,t,y, and a to be symbols
f    = x + sin(x)         % define f(x)
q    = 3*t^2 -7^t         % build up a horrible formula
g    = subs(f,x,q )       % Substitute q(t) in for x
h    = subs(g,t, 'exp(y/a)')  % Substitute exp(y/a) in for t
pretty(h)                 % Print it in a readable form

result = subs(h,{y,a},{7,9})
                          % Evaluate it with y=7 and a=9
a_number_please = double(result)
                          % Get a number you can work with!

y=0:.1:1;  a = pi;        % Try an array of values -
                          % for y and a
y=sym(y); a=sym(a);       % treat the values of -
                          % y and a as symbols

hvec   = vectorize(h)     % Write a formula that works -
                          % with arrays
result = eval(hvec)'      % Evaluate that vectorized -
                          % formula (mess!)

result_numeric =  double(result)   % Maple's exact --> numbers
plot (double(y), result_numeric)   % Graph the horrible formula
```

Comments:

- Using substitution, it is easy to build up big messy formulas like the formula stored in h.
- The command pretty sometimes can help you see through a messy formula. Try also simplify to reduce formulas using common trig identities, collect common terms, and so on.

- We have used a fancier syntax for **subs** here by substituting two things at once.
- Maple writes formulas in a reasonable way for functions of one variable. MATLAB is set up for matrices. To get MATLAB to plug in a formula for an array of numbers (and not perform matrix operations) you have to use the funny 'dot' syntax (like `y.^2` to square all the elements of `y`). The **vectorize** command takes a formula that is good for scalars and puts dots in the right places.
- When to use (or not) the **double** or **eval** commands is perfectly confusing. Trial and error will be an inevitable part of your work.

3. **Make an inline function.** Sometimes it is convenient to have a new "function" to work with but you don't want to write a whole M-file for the purpose. You would like to be able to type `myfun(7)` and have a big formula evaluated. In particular, you might like this formula to be one you cooked up with the Symbolic toolbox.

Say, you want to know how one of the roots of a cubic polynomial depends on one of the coefficients. Here is one approach.

```
syms x a
f = x^3 + a* x^2 + 3*x +5    % A cubic  with parameter a.
roots = solve(f,x)           % Find the three roots (a mess!).
root1 = roots(1)             % Pick out the first root (a mess!).

myfun=inline(char(root1))    % Make an inline function.
myfun(7)                     % Find the root when a=7.
```

Comments:

- `root1` is the symbolic expression for the first root of the cubic polynomial in terms of the parameter `a`.
- The **inline** function wants a character (string) expression not a symbolic expression (even though they look the same when typed out) so you have to convert the expression using the **char** function.
- If you want to plug in a list of values for `a` all at one time you can change the last two lines as follows:

```
myfun = inline( char(vectorize(root1)) )
myfun(4:.2:8)'           % a, from 4 to 8.
```

8.1.5   Summary: some Symbolic Math Toolbox commands

Here is a list of the commands we have discussed (and one more). The things inside the ()'s are there as examples. You could change them.

`syms x t y a b`	declare x, t, y, a, b as symbolic. Do this first.
`sym([1 2 3])`	Treat the list [1 2 3] as symbolic.
`diff(sin(t),t)`	Calculate $\frac{d}{dt}\sin t$ to be $\cos t$
`int(x^2, x)`	Calculate $\int x^2\, dx$ to be $x^3/3$
`expand((a+b)^3)`	Expand $(a+b)^3$ to $a^3 + 3a^2b + 3ab^2 + b^3$
`collect(x^2 + 2*x^2)`	Collect $x^2 + 2x^2$ into $3x^2$
`solve(x^3-y, x)`	Find the 3 cube roots of y.
`subs(x*y^2, y, a)`	Substitute $y = a$ into xy^2 to get xa^2.
`pretty(x^2)`	Print out x^2 on two or more easy to read lines.
`simplify(sin(t)^2 +cos(t)^2)`	Use trig to simplify $\sin^2 t + \cos^2 t$ to 1.
`eval('sin(7)')`	Treat the string `'sin(7)'` as a MATLAB command.
`double(sym(rand(3)))`	Get numbers from a symbolic array.
`vectorize('x^2')`	Put dots in an expression (turn `x^2` into `x.^2`).
`char(sym(t^2))`	Turn a symbolic expression into a char string.
`f=inline('x^2-x')`	Make `f` so that, say, `f(7)` calculates $7^2 - 7$.
`ezplot(exp(t))`	The simplest way to plot e^t.
`funtool`	Bring up a symbolic graphing calculator.

There are mainly three other significant facilities in MATLAB, which we do not discuss in depth here.

8.2   Debugging Tools

MATLAB supports a built-in debugger which consists of several commands — `dbclear`, `dbcont`, `dbdown`, `dbquit`, `dbstack`, `dbstatus`, `dbstep`, `dbstop`, `dbtype`, and `dbup`, to help you debug your MATLAB programs. You can write these commands in your M-file that you want to debug, or you can invoke them interactively by clicking on them in the Editor/Debugger window. See the on-line help under `debug` for details of these commands.

8.3   External Interface: Mex-files

If you wish to dynamically link your Fortran or C programs to MATLAB functions so that they can communicate and exchange data, you need to learn about *MEX*-files. Consult the MATLAB Application Program Interface Guide [3] to learn about these files. The process of developing Mex-files is fairly complicated and highly system-dependent. You should perhaps first consider non-dynamic linking with your external programs through standard ASCII data files.

8.4   Graphics User Interface

It is also possible to design your own Graphical User Interface (GUI) with menus, buttons, and slider controls in MATLAB. This facility is very useful if you are developing an application package to be used by others. You can build many visual 'user-friendly' features in your application. For more information consult Building GUIs with MATLAB [4].

A. The MATLAB Language Reference

A.1 Punctuation Marks and Other Symbols

For on-line help type:
`help punct`

, **Comma:** A comma is used to:
  - separate variables in the input and output list of a function,
    Example: `[t,x]=ode23('pend',t0,tf,x0),`
  - separate the row and column indices in a matrix,
    Example: `A(m,n), A(1:10,3)` etc.,
  - separate different commands on the same line.
    Example: `plot(x,y), grid, xlabel('x')` etc.

; **Semicolon:** A semicolon is used to:
  - suppress the MATLAB output of a command,
    Example: `x=1:10; y=A*x;` etc.,
  - separate rows in the input list of a matrix.
    Example: `A=[1 2; 4 9].`

: **Colon:** A colon is used to specify range:
  - in creating vectors,
    Example: `x=1:10; y=1:2:100;` etc.
  - for matrix and vector indices,
    Example: see Section 3.1.3,
  - in `for` loops.
    Example: `for i=1:20, x=x+i; end.`

' **Right Quote:** A single right quote is used to transpose a vector or a matrix.
    Example: `symA = (A'+A)/2.`

' ' **Single Quotes:** A pair of single right quote characters is used to enclose a character string.

Example: `xlabel('time'), title('My plot')` etc.

. **Period:** A period is used:
- as a decimal point,
- in array operations.

Example: `Asq = A.^2` (see page 58).

.. **Two periods:** Two periods are used in `cd ..` command to access parent directory.

... **Ellipsis:** Ellipsis (three periods) at the end of a command denote continuation to the next line.

Example: `x = [log(1:100) sin(v+a.*b) 22.3 23.0 ...`
`        34.0 33.0 40:50 80];`

! **Exclamation:** An exclamation preceding a *command* is used to send the local operating system command *command* to the system. This command is not applicable to Macs.

Example: `!emacs newfile.m` invokes the local emacs editor.

% **Percent:** A percent character is used to:
- used to mark the beginning of a comment, except when used in character strings.

Example: `% This is a comment`, but `rate = '8.5%'` is a string,
- to denote formats in standard I/O functions `sprintf` and `fprintf`.

Example: `sprintf('R = %6.4f', r)`.

( ) **Parentheses:** Parentheses are used to:
- specify precedence in arithmetic operations,

Example: `a = 5/(2+x*(3-i));` etc.
- enclose matrix and vector indices,

Example: `A(1:5,2) = 5; v = x(1:n-5);` etc.
- enclose the list of input variables of a function.

Example: `[t,x]=ode23('pend', t0, tf, x0)`.

[ ] **Square brackets:** Square brackets are used to:
- form and concatenate vectors and matrices,

Example: `v = [1 2 3:9]; X = [v; log(v)];` etc.
- enclose the list of output variables of a function.

Example: `[V,D] = eig(A);` etc.

A.2   General-Purpose Commands

See Section 1.6.6

For on-line help type:
`help general`

Help & Query			
`lookfor`	Keyword search for help	`whatsnew`	Display ReadMe files
`help`	On-line help	`what`	List files in the directory
`demo`	Run demo program	`which`	Locate a file
`info`	Info about MATLAB	`why`	Give philosophic advice
`ver`	MATLAB version info	`path`	List accessible directories

Command Window Control			
`clc`	Clear command window	`home`	Send cursor home
`format`	Set screen output format	`echo`	Echo commands in script file
`more`	Control paged screen output	$\uparrow, \downarrow$	Recall previous commands

Working with Files & Directories			
`pwd`	Show current directory	`delete`	Delete file
`cd`	Change current directory	`diary`	Save text of MATLAB session
`dir, ls`	List directory contents	`type`	Show contents of file
`mkdir`	Create a new directory	`!`	Access operating system

Variable and Workspace			
`clear`	Clear variables and functions	`length`	Length of a vector
`who,whos`	List current variables	`size`	Size of a matrix
`load`	Load variables from file	`pack`	Consolidate memory space
`save`	Save variables in MAT-file	`disp`	Display text or matrix

Start & Exit			
`matlabrc`	Master startup file	`quit`	Quit MATLAB
`startup`	M-file executed at startup	`exit`	Same as quit

Time & Date			
`clock`	Wall clock time	`etime`	Elapsed time function
`cputime`	Elapsed CPU time	`tic`	Start stopwatch timer
`date`	Date, month, year	`toc`	Read stopwatch timer

A.3 Special Variables and Constants

Constants		Variables	
pi	π (=3.14159...)	ans	Default output variable
inf	∞ (infinity)	computer	Computer type
NaN	Not-a-Number	nargin	Number of input arguments
i, j	Imaginary unit ($\sqrt{-1}$)	nargout	Number of output arguments
eps	Machine precision		
realmax	Largest real number		
realmin	Smallest real number		

For on-line help type:
`help lang`

A.4 Language Constructs and Debugging

See Section 4.3.

Declarations/Definitions		
script	function	global
nargchk	persistent	mlock

Interactive Input Functions		
input	keyboard	uimenu
ginput	pause	uicontrol

Control Flow Functions		
for	while	end
if	elseif	else
switch	case	otherwise
error	break	return

Debugging				
dbclear	dbcont	dbstep	dbstack	dbstatus
dbup	dbdown	dbtype	dbstop	dbquit

For on-line help type:
`help iofun`

A.5 File I/O

See Section 4.3.7.

File Opening, Closing, and Positioning					
fopen	fclose	fseek	ftell	frewind	ferror

File Reading and Writing					
fread	fwrite	fprintf	fscanf	fgetl	fgets

A.6 Operators and Logical Functions

See Section 3.2.

For on-line help type:
`help ops`

Arithmetic Operators			
Matrix Operators		*Array Operators*	
+	Addition	+	Addition
−	Subtraction	−	Subtraction
*	Multiplication	.*	Array multiplication
^	Exponentiation	.^	Array exponentiation
/	Left division	./	Array left division
\	Right division	.\	Array right division.

Relational Operators		Logical Operators	
<	Less than	&	Logical AND
<=	Less than or equal	|	Logical OR
>	Greater than	~	Logical NOT
>=	Greater than or equal	xor	Logical EXCLUSIVE OR
==	Equal		
~=	Not equal		

Logical Functions			
all	any	exist	find
finite	isempty	isinf	isnan
isiee	issparse	isstr	isfinite

A.7   Math Functions

For on-line help
type:
`help elfun`

See Section 3.2.4 for description and examples.

Trigonometric Functions			
sin	asin	sinh	asinh
cos	acos	cosh	acosh
tan	atan,atan2	tanh	atanh
cot	acot	coth	acoth
sec	asec	sech	asech
csc	acsc	csch	acsch

Exponential Functions			
exp	log	log10	sqrt
log2	pow2	nextpow2	

For on-line help
type:
`help specfun`

Complex Functions			
abs	angle	conj	complex
real	imag	unwrap	cplxpair

Round-off Functions			
fix	floor	ceil	round
rem	sign	mod	

Specialized Math Functions			
bessel	bessely	besselh	beta
betain	betaln	ellipj	ellipke
erf	erfinv	gamma	gammainc
legendre	rat	dot	cross

A.8 Matrices: Creation & Manipulation

See Section 3.1.

For on-line help type:
`help elmat`

Elementary Matrices			
eye	ones	zeros	rand
randn	linspace	logspace	meshgrid

Specialized Matrices			
compan	hadamard	hankel	hilb
invhilb	magic	pascal	rosser
toeplitz	vander	wilkinson	gallery

Matrix Manipulation Functions			
diag	fliplr	flipud	reshape
rot90	tril	triu	:

For on-line help type:
`help matfun`

Matrix (Math) Functions			
expm	logm	sqrtm	funm

Matrix Analysis			
cond	det	norm	null
orth	rank	rref	trace
eig	balance	poly	hess

Matrix Factorization & Inversion			
chol	cholinc	lu	luinc
eig	eigs	svd	svds
qr	qz	schur	pinv

Sparse Matrix Functions There are also several functions for creating, manipulating, and visualizing sparse matrices. Some of these are spdiag, speye, sprandn, full, sparse, spconvert, spalloc, spfun, condest, normest, sprank, gplot and spy. See on-line help for complete listing.

For on-line help type:
`help sparfun`

A.9   Character String Functions

For on-line help
type:
help strfun

See Section 3.2.6.

General String Functions					
abs	char	eval	setstr	strcat	strvcat
string	strcmp	lower	upper	isstr	ischar

String $\Longleftrightarrow$ Number Conversion				
int2str	num2str	sprintf	dec2hex	mat2str
str2num	sscanf	hex2dec	hex2num	dec2bin

A.10   Graphics Functions

For on-line help
type:
help graphics
help graph2d
help graph3d

See Chapter 6.

EZ Graphics				
ezplot	ezpolar	ezcontour	ezcontourf	ezgraph3
ezplot3	ezmesh	ezmeshc	ezsurf	ezsurfc

2-D Graphics				
plot	loglog	semilogx	semilogy	fplot
bar	errorbar	compass	feather	stairs
polar	fill	hist	rose	quiver

3-D Graphics				
plot3	fill3	mesh	meshc	meshz
surf	surfc	surfl	cylinder	sphere

Contour Plots				
contour	contour3	contourc	clabel	pcolor

Graphics Annotation				
xlabel	ylabel	zlabel	title	legend
text	gtext	grid	plotedit	

Axis Control & Graph Appearance				
axis	colormap	hidden	shading	view

Window Creation & Control				
clf	close	figure	gcf	subplot

Axis Creation & Control				
axes	axis	caxis	cla	gca

Handle Graphics Objects & Operations				
axes	line	patch	surface	text
figure	image	uicontrol	uimenu	
delete	drawnow	get	reset	set

Animation & Movies				
comet	capture	getframe	movie	moviein
frame2im	im2frame	rotate		

Hardcopy & Miscellaneous				
print	orient	printopt	ginput	hold

Color Control & Lighting				
caxis	colormap	flag	hsv2rgb	rgb2hsv
bone	copper	gray	hsv	pink
cool	hot	shading	brighten	diffuse
surfl	specular	rgbplot		

For on-line help type:
`help color`

A.11   Applications Functions

For on-line help type:
`help datafun`

A.11.1   Data analysis and Fourier transforms

Basic Statistics Commands				
mean	median	std	min	max
prod	cumprod	sum	cumsum	sort

Correlation & Finite Difference				
corrcoef	cov	del2	diff	gradient

Fourier Transforms				
fft	fft2	fftshift	ifft	ifft2
abs	angle	cplxpair	nextpow2	unwrap

Filtering & Convolution				
conv	conv2	dconv	filter	filter2

For on-line help type:
`help polyfun`

A.11.2   Polynomials and data interpolation

Polynomials				
poly	polyder	polyfit	polyval	polyvalm
conv	deconv	residue	roots	

Data Interpolation			
interp1	interp2	interpft	griddata

Fourier Transforms				
fft	fft2	fftshift	ifft	ifft2
abs	angle	cplxpair	nextpow2	unwrap

Filtering & Convolution				
conv	conv2	dconv	filter	filter2

For on-line help type:
`help funfun`

A.11.3   Nonlinear numerical methods

Functions					
fmin	fmins	fminbnd	fminsearch	fzero	trapz
quad	quadl	dblquad	bvp4c	pdepe	pdeval
ode23	ode45	ode113	ode23t	ode23s	odefile

Bibliography

[1] *Using MATLAB, Version 6*, The MathWorks, Inc., 2000.

[2] *MATLAB Function Reference*, The MathWorks, Inc., 2000.

[3] *MATLAB Application Program Interface Reference, Version 6*, The MathWorks, Inc., 2000.

[4] *Creating Graphical User Interfaces, Version 1*, The MathWorks, Inc., 2000.

[5] *Using MATLAB Graphics, Version 6*, The MathWorks, Inc., 2000.

[6] *MATLAB Release Notes for Release 12*, The MathWorks, Inc., 2000.

[7] Kernighan, B. W., and D. M. Ritchie, *The C Programming Language*, second edition, Prentice-Hall Inc., 1993.

[8] Strang, Gilbert, *Linear Algebra and Its Applications*, third edition, Saunders HBJ College Publishers, 1988.

[9] Golub, G. H., and C. F. Van Loan, *Matrix Computations*, third edition, The Johns Hopkins University Press, 1997.

[10] Horn, R. A., and C. R. Johnson, *Matrix Analysis*, Cambridge University Press, 1985.

[11] Gerald, C. F., and P. O. Wheatley, *Applied Numerical Analysis*, fifth edition, Addison Wesley Publishing Company, 1994.

[12] Press W., B. Flannery, S. Teudolsky, and W. Vetterling, *Numerical Recipes in C: The Art of Scientific Computing*, second edition, Cambridge University Press, 1992.

Index